Transport & Fate of Chemicals in Soils

Principles & Applications

Transport & Fate
of Chemicals in Soils

Principles & Applications

H. Magdi Selim

CRC Press
Taylor & Francis Group
Boca Raton London New York

CRC Press is an imprint of the
Taylor & Francis Group, an **informa** business

CRC Press
Taylor & Francis Group
6000 Broken Sound Parkway NW, Suite 300
Boca Raton, FL 33487-2742

First issued in paperback 2017

© 2015 by Taylor & Francis Group, LLC
CRC Press is an imprint of Taylor & Francis Group, an Informa business

No claim to original U.S. Government works

ISBN-13: 978-1-4665-5794-9 (hbk)
ISBN-13: 978-1-138-07592-4 (pbk)

Visit the Taylor & Francis Web site at
http://www.taylorandfrancis.com

and the CRC Press Web site at
http://www.crcpress.com

To my wife and son, Liz and Richard

Contents

Preface

Over the last four decades tremendous advances have been made toward the understanding of transport characteristics of the transport contaminants in soils, as well as solutes and tracers in geological media. These advances were broad in nature and addressed physical and chemical processes that influence the behavior of solutes in porous media. Examples include nonlinear kinetics, release and desorption hysteresis, multisite and multireaction reactions, and competitive-type reactions. Examples that focus on physical processes include fractured media, multiregion, multiple porosities, and heterogeneity and effect of scale.

This book provides the basic framework of the principles governing the sorption and transport of chemicals in soils. The physical and chemical processes presented above are the focus of this textbook. Details of sorption behavior of chemicals with soil matrix surfaces are presented, as well the integration of sorption characteristics with mechanisms that govern solute transport in soils. Applications of the principles of sorption and transport are not restricted to chemical contaminants. Rather, examples include tracers, phosphorus, and trace elements, including essential micronutrients, heavy metals, military explosives, pesticides, and radionuclides.

This book should be a useful textbook for undergraduate and graduate students in soil science, environmental science and engineering, chemical and civil engineering, hydrology, and geology. It should also be useful to scientists, engineers, and geologists, and to federal and state agencies, researchers, consulting engineers, and decision makers in the management and restoration of contaminated sites.

In Chapter 1, the concept of sorption isotherms, their classification, and modeling approaches are discussed. The effect of soil properties and differences of empirical and mechanistic approaches in describing isotherm relations are presented. Chapter 2 is devoted to describing kinetic approaches for solute sorption and the limitations of equilibrium retention concepts. Adsorption–desorption hysteresis is presented and evidence of kinetics is discussed. Laboratory methods are presented for obtaining kinetic data for adsorption and desorption.

In Chapters 3 and 4, principles of solute transport and mathematical models as a tool for the description of the retention reactions of solutes during transport in the soil are presented. Analytical solutions of the convection–dispersion equation, for various conditions, are presented in Chapter 3. For cases where solute behavior is nonlinear, numerical solutions are discussed. In Chapter 4, the dependence of dispersivity on the mean travel distance and/or scale of the geologic system is discussed.

In Chapter 5, two-site and multireaction models that describe multiple interactions of solutes in soils are presented. This is followed by second-order models as discussed in Chapter 6. Furthermore, implications of equilibrium and second-order formulations and the kinetic Langmuir equation are discussed. Competitive behavior of solutes in soils is the focus of Chapter 7. Examples are presented, followed by a discussion of equilibrium ion exchange models and modified kinetic ion exchange and applications. Several empirical formulations for competitive sorption are also presented. Competitive Freundlich approaches for multiple ions are discussed and modified competitive kinetics presented.

The last two chapters are devoted to examining dominant nonequilibrium processes for transport of solute in soils. Physical heterogeneity attributes are discussed in the framework of nonuniformity of pore space, soil aggregation, soil layering, and cavities developed from natural and anthropogenic activities. In Chapter 8, we discuss physical heterogeneity, with emphasis on matrix diffusion, preferential flow, mobile–immobile, two-region, and multidomain flow. In Chapter 9, emphasis is on other sources of soil heterogeneity. Transport in nonuniform media due to soil layering or stratification is discussed as an intrinsic part of soil formation processes. Also discussed in Chapter 9 are other sources of soil heterogeneity and their influence on the behavior of solute transport in soils. This includes nonuniform media, often as a consequence of mixing of different soils and geological media during industrial and agricultural activities.

I wish to thank various individuals for their assistance in completing this book. I owe special thanks to Robert Mansell, the University of Florida; Alex Iskandar, the U.S. Army Cold Region Research and Engineering Laboratory; Mike Amacher, U.S. Department of Agriculture (USDA) Forest Service; Logan Utah, Hannes Flühler, and Rainer Schulin, ETH, Zurich, Switzerland; and Donald Sparks, the University of Delaware. Their contributions to my soil physics program at Louisiana State University were valuable and extremely rewarding. I also acknowledge my current and former graduate students and visiting scientists who made significant contributions to the advancement of our present understanding of solute behavior in soils. Special thanks go to Christoph Hinz, the University of Western Australia, Perth; Hua Zhang, Chinese Academy of Sciences, Yanti, China; Tamer Elbana, National Research Center, Cairo, Egypt; Bernhard Buchter, ETH, Zurich, Switzerland; Glenn Wilson, USDA Agricultural Research Service Sedimentation Laboratory, Oxford, Mississippi; Lixia Liao, U.S. Geological Survey, Palo Alto, California; Steven Lynn McGowen, USDA National Resources Conservation Service, Oklahoma City; Liwang Ma, Modeling Unit, USDA, Ft. Collins, Colorado; Hongxia Zhu, SIAC, Reston, Virginia; Liuzong Zhou; and Eric Ferguson. Special thanks go to Brent Clothier for providing the original of Figure 8.1.

This book would not have been possible without the continued support of the faculty, staff, and administration of the Louisiana State University. Finally special thanks also go to the editorial staff of Taylor and Francis/CRC Press, in particular Irma Britton, for their suggestions and efforts in the publication of this book.

Magdi Selim

The Author

H. Magdi Selim is professor of soil physics and George and Mildred Caldwell Endowed Professor of Soil Science, School of Plant, Environmental and Soil Sciences at Louisiana State University at Baton Rouge. He earned his BS degree in soil science from Alexandria University, Alexandria, Egypt, and his MS and PhD in soil physics from Iowa State University, Ames, Iowa. Professor Selim is internationally recognized for his research in the areas of the transport and kinetics of reactive chemicals in heterogeneous porous media and modeling of dissolved chemicals in water-saturated and unsaturated soils. He is the original developer of several conceptual models for describing the retention processes of dissolved chemicals in soils and natural materials in porous media. His pioneering work also includes two-site and multireaction and nonlinear kinetic models for trace elements, heavy metals, radionuclides, explosive contaminants, phosphorus, and pesticides in soils and geological media.

Professor Selim is the author or coauthor of numerous scientific publications in several journals, and is author or coauthor of several books. He is a member of the American Society of Agronomy, Soil Science Society of America, International Society of Soil Science, International Society of Trace Element Biogeochemistry, Louisiana Association of Agronomy, and American Society of Sugarcane Technology. Dr. Selim was elected chair of the Soil Physics Division of the Soil Science Society of America. He has served on numerous committees of the Soil Science Society of America, the American Society of Agronomy, and the International Society of Trace Element Biogeochemistry. He has also served as associate editor of *Water Resources Research* and the *Soil Science Society of America Journal*, and as technical editor of the *Journal of Environmental Quality*.

Professor Selim is a Fellow of the American Society of Agronomy and the Soil Science Society of America, and is the recipient of several professional awards and honors, including Phi Kappa Phi Research, Gamma Sigma Delta Award for Research, Joe Sedberry Graduate Teaching Award, First Mississippi Research Award, Doyle Chambers Achievements Award, and EPA Environmental Excellence Award. Recent awards include the Soil Science Research Award from the Soil Science Society of America (SSSA) and the von Liebeg Award from the International Union of Soil Science (IUSS).

1

Sorption Isotherms

Adsorption isotherms or, more accurately, sorption isotherms are convenient ways to graphically represent the amount of an adsorbed compound, or adsorbate, in relation to its concentration in the equilibrium solution or adsorbent. In other words an adsorption isotherm is a relationship relating the concentration of a solute on the surface of an adsorbent to the concentration of the solute in the liquid with which it is in contact at a constant temperature. Freundlich and Langmuir sorption isotherms are extensively used to describe sorption isotherms for a wide range of chemicals. Knowledge of sorption isotherms and adsorption phenomena is essential for understanding heavy solute retention and transport in soils and geological media. It is crucial for assessing the environmental risk of contamination and pollution provoked by these elements. Studies on solutes adsorption in soils are often conducted as one-component systems, where the ions or molecules are treated individually, or they can be conducted as a multicomponent system, where the ions are subjected to competition among them.

1.1 Freundlich Isotherm

The Freundlich isotherm is one of the simplest approaches for quantifying the behavior of retention of reactive solute with the matrix surfaces. It is certainly one of the oldest nonlinear sorption equations and was based on quantifying the gas adsorbed by a unit mass of solid with pressure, which was described as

$$S = K_f C^b \qquad (1.1)$$

where S is the amount of solute retained by the soil in $\mu g\ g^{-1}$ or mg kg^{-1}, C is the solute concentration in solution in mg L^{-1} or μg ml^{-1}, K_f is the partitioning coefficient in L kg^{-1} or ml g^{-1}, and b is often considered a dimensionless parameter. A major disadvantage of the Freundlich approach is that it is incapable of describing sorption at or near saturation or when the adsorption maximum is reached. The parameter K_f represents the partitioning of a solute species between solid and liquid phases over the concentration range of interest and is analogous to the equilibrium constant for a chemical reaction. For b equal unity, the Freundlich equation takes on the form of the linear equation:

$$S = K_d C \qquad (1.2)$$

where K_d is the distribution coefficient in L/kg or ml/g. As K_d is used, this implies a linear, zero-intercept relationship between sorbed and solution concentration, which is a convenient assumption but certainly not universally true. This linear model is often referred to as a constant partition model and the parameter K_d is a universally accepted environment parameter that reflects the affinity of matrix surfaces to solute species. The K_d parameter provides an estimate of the potential for the adsorption of dissolved contaminants in contact with soil. It is typically used in fate and contaminant transport calculations. According to the U.S. Environmental Protection Agency (1999), K_d is defined as the ratio of the contaminant concentration associated with the solid to the contaminant concentration in the surrounding aqueous solution when the system is at equilibrium.

1.2 Langmuir Isotherm

The Langmuir isotherm was developed to describe the adsorption of gases by solids where a finite number of adsorption sites in planer surfaces is assumed (Langmuir, 1918). Major assumptions include that ions are adsorbed as a monolayer on the surface, and the maximum adsorption occurs when the surface is completely covered. Other assumptions are that the surface considered is homogeneous, the sites are of identical adsorption energy or affinity over the entire surface, and that equilibrium condition is attained.

As a result, a major advantage of the Langmuir equation over the linear and Freundlich types is that a maximum sorption capacity is incorporated into the formulation of the model, which may be regarded as a measure of the amount of available retention sites on the solid phase. The standard form of the Langmuir equation is

$$\frac{S}{S_{max}} = \frac{\omega C}{1 + \omega C} \qquad (1.3)$$

where ω and S_{max} are adjustable parameters. Here ω (ml μg^{-1}) is a measure of the bond strength of molecules on the matrix surface and S_{max} ($\mu g\ g^{-1}$ of soil) is the maximum sorption capacity or total amount of available sites per unit soil mass.

1.2.1 Two-Site Langmuir

Based on several retention data sets, the presence of two types of surface sites responsible for the sorption of phosphate (P) in several soils was postulated. As a consequence, the Langmuir two-surface isotherm was proposed (Holford, Wedderburn, and Mattingly, 1974) such that:

$$\frac{S}{S_{max}} = F_1\left(\frac{\omega_1 C}{1 + \omega_1 C}\right) + F_2\left(\frac{\omega_2 C}{1 + \omega_2 C}\right) \tag{1.4}$$

where F_1 and F_2 (dimensionless) are considered as fractions of type 1 and type 2 sites to the total sites ($F_1 + F_2 = 1$), and ω_1 and ω_2 are the Langmuir coefficients associated with sites 1 and 2, respectively. Equation 1.4 is an adaptation of the original equation proposed by Holford, Wedderburn, and Mattingly (1974) and was used by Holford and Mattingly (1975) to describe P isotherms for a wide range of soils.

A more recent adaptation of the two-surface Langmuir equation is the incorporation of a sigmoidicity term, where

$$\frac{S}{S_{max}} = F_1\left(\frac{\omega_1 C}{1 + \omega_1 C + (\sigma_1/C)}\right) + F_2\left(\frac{\omega_2 C}{1 + \omega_2 C + (\sigma_2/C)}\right). \tag{1.5}$$

The terms σ_1 and σ_2 are the sigmodicity coefficients ($\mu g\ cm^{-3}$) for type 1 and type 2 sites, respectively. Schmidt and Sticher (1986) found that the introduction of this sigmoidicity term was desirable in order to adequately describe sorption isotherms at extremely low concentrations. Although the Langmuir approach has been used to model P retention and transport from renovated wastewater, the two-surface Langmuir with sigmodicity has been rarely used to describe heavy metal retention during transport in soils. It should be emphasized that for both Langmuir and Freundlich isotherms, an implicit assumption is always made, namely, that equilibrium conditions are dominant. Therefore, these formulations do not take into consideration possible time-dependent or kinetic sorption behavior. Such sorption behavior is common for most solutes and is considered in Chapter 2.

1.3 General Types of Isotherms

Giles, D'Silva, and Easton (1974) proposed four general types of isotherms (C, L, H, and S), which are illustrated in Figure 1.1. The C-type isotherms indicate partitioning of ions or molecules between the solution and sorbed phase along the same lines as that of the linear model (Equation 1.2). The L-type

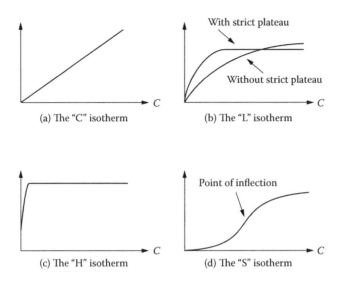

FIGURE 1.1
The four main types of isotherms. (After Giles, D'Silva, and Easton, 1974.)

isotherm is characterized by decreasing slopes as the vacant sites become occupied by the sorbed ion or molecule. Freundlich and Langmuir isotherms are commonly referred to as L-curve isotherms. In L-type isotherms, at low solution concentrations, high-energy sites are occupied first. Subsequently, as the concentration in solution increases, sites of moderate and low affinities become occupied. The H-type isotherms are best characterized by extremely high sorption possibly due to irreversible reactions. The S-type isotherms indicate low affinity for sorption at low solution concentrations followed by a gradual sorption increase. At higher concentration, sorption decreases and sorption maximum is perhaps attained.

In the literature, L-type isotherms are frequently encountered for most trace elements and for heavy metals. Specifically, Freundlich and Langmuir isotherms are adopted for a wide range of solutes. C-type isotherms are often observed for pesticides (herbicides and insecticides) where linear isotherms are often observed.

1.4 Empirical versus Mechanistic Models

Fontes (2012) argued that the adsorption phenomenon can be represented by two main conceptual models: empirical models initially derived from experiments, and semiempirical or mechanistic models, based on reaction mechanisms. The main difference between these two types of models is the absence of an electrostatic term in the empirical models, whereas it is mandatory in mechanistic models. Mechanistic models, also known as chemical models, are expected to provide a close representation of the real adsorption phenomenon in the soil system. Nevertheless, due to the complexity of chemical model, empirical models are utilized in most solute studies with soils and geological media. Moreover, empirical models have been widely used in soil science and environmental studies related to metal and anion adsorption and pesticide retention in soils. A listing of such models is presented in Table 1.1. These models do not take into consideration the electrostatic influence of the electrically charged surfaces in the solution, or the influence of changes in surface charges due to the composition of the soil solution. In the empirical model, the model form is chosen a posteriori from the observed adsorption data and to enable a satisfying fitting of the experimental data

TABLE 1.1

Selected Equilibrium and Kinetic Type Models for Heavy Metal Retention in Soils

Model	Formulation[a]
Equilibrium Type	
Linear	$S = K_d C$
Freundlich	$S = K_f C^b$
General Freundlich	$S/S_{max} = [\omega C/(1 + \omega C)]^\beta$
Rothmund-Kornfeld ion exchange	$S_i/S_T = K_{RK} (C_i/C_T)^n$
Langmuir	$S/S_{max} = \omega C/[1 + \omega C]$
General Langmuir Freundlich	$S/S_{max} = (\omega C)^\beta/[1 + (\omega C)^\beta]$
Langmuir with Sigmoidicity	$S/S_{max} = \omega C/[1 + \omega C + \sigma/C]$
Kinetic Type	
First-order	$\partial S/\partial t = k_f (\theta/\rho)C - k_b S$
*n*th order	$\partial S/\partial t = k_f (\theta/\rho)C^n - k_b S$
Irreversible (sink/source)	$\partial S/\partial t = k_s(\theta/\rho) (C - C_p)$
Second-order irreversible	$\partial S/\partial t = k_s(\theta/\rho) C (S_{max} - S)$
Langmuir kinetic	$\partial S/\partial t = k_f (\theta/\rho) C(S_{max} - S) - k_b S$
Elovich	$\partial S/\partial T = A\text{Exp}(-Bs)$
Power	$\partial S/\partial t = K(\theta/\rho) C^n S^m$
Mass transfer	$\partial S/\partial t = K(\theta/\rho) (C - C^*)$

[a] $A, B, b, C^*, C_p, K, K_d, K_{RK}, k_b, k_f, k_s, n, m, S_{max}, \omega, \beta$, and σ are adjustable model parameters, ρ is bulk density, Θ is volumetric soil water content, C_T is total solute concentration, and S_T is total amount sorbed of all competing species.

the mathematical form and the number of parameters are chosen to be as simple as possible (Bradl, 2004).

Surface-complexation models (SCMs) are chemical models based on a molecular description of the electric double layer using equilibrium-derived adsorption data. To apply SCMs, it is required to provide the amount of reactive surface of each type of sorbent as well as the density of the reaction sites. In reality, experimental characterization of soil minerals is time consuming and data is rarely available. In addition, the application of SCMs requires the estimation of an extensive list of specific surface species and their thermodynamic reaction constants. Moreover, basic to SCMs such as MINTEQA2 and PHREEQC is that equilibrium conditions are assumed dominant, that is, LEA (local equilibrium assumption) is valid. The drawback of SCM models is that they are limited to describing equilibrium-type reactions and do not account for kinetic sorption–desorption processes in soils. In heterogeneous soils with a variety of sorbents having different reactivities, time-dependent sorption is often observed. Kinetics or nonequilibrium sorption may arise due to the heterogeneity of sorption sites on matrix surfaces. In fact, various types of surface complexes (e.g., inner sphere, outer sphere, monodentate, bidentate, mononuclear, or binuclear) with contrasting sorption affinities can be formed on mineral surfaces with metals and metalloids. This heterogeneity of sorption sites may contribute to the observed adsorption kinetics where sorption takes place preferentially on high-affinity sites, followed subsequently by slow sorption on sites with low sorption affinity. Furthermore, the diffusion of ions to reaction sites within the soil matrix has been proposed as an explanation for time-dependent adsorption by many researchers (e.g., Fuller, Davis, and Waychunas, 1993; Raven, Jain, and Loeppert, 1998). The PHREEQC model simulates sorption kinetics with first-order kinetic equations. PHREEQC was not developed to simulate nonlinear adsorption kinetics. Recent work of Zhang (2013) represents efforts to account for kinetics by relaxing the equilibrium assumption in SCMs.

Surface-complexation models have been used to describe an array of equilibrium-type chemical reactions, including proton dissociation, metal cation and anion adsorption reactions on oxides and clays, organic ligands adsorption, and competitive adsorption reactions on oxide and oxide-like surfaces. Application and theoretical aspects of SCMs are extensively reviewed by Goldberg (1992) and Sparks (2003). SCMs are chemical models based on molecular description of the electric double layer using equilibrium-derived adsorption data. They include the constant capacitance model, triple-layer model, Stern variable surface charge models, among others. SCMs have been incorporated into various chemical speciation models. The model MINEQL was perhaps the first where the chemical speciation was added to the triple-layer SCM. Others include MINTEQ, SOILCHEM, HYDRAQL, MICROQL, and FITEQL (see Goldberg, 1992).

All these chemical equilibrium models require knowledge of the reactions involved and associated thermodynamic equilibrium constants. Due to the heterogeneous nature of soils, extensive laboratory studies may be needed to determine these reactions. Thus, predictions from transport models based on the surface-complexation approach may not describe heavy metal sorption by a complex soil system. As a result, the need for direct measurements of the sorption and desorption/release behavior of heavy metals in soils is necessary. Consequently, retention, or the commonly used term sorption, should be used when the mechanism of heavy metals removal from soil solution is not known, and the term adsorption should be reserved for describing the formation of solute-surface site complexes.

1.5 Nonlinearity and Heterogeneity

The description of sorption isotherms in soils remains empirical where the Freundlich and Langmuir models are commonly used. For several decades, however, it has been recognized that isotherm patterns or the shape of an adsorption isotherm is a reflection of the heterogeneity of the soil matrix. The fact that a soil is made up of numerous constituents with distinctly different properties lends credence to this general concept. One may view a soil as a complex mixture of numerous constituents, thus forming a highly heterogeneous system. Consequently, the following general isotherm equation that describes the affinity of solutes to different sorption sites on surfaces of soils was proposed:

$$S = \int g(\zeta)\Gamma(\zeta, C)d\zeta \tag{1.6}$$

where $g(\zeta)$ represents the affinity distribution of a soil for a specific chemical or may be referred to as a probability distribution function (PDF; Kinnebergh, 1986). The function $\Gamma(\zeta, C)$ represents the sorption isotherm function used to represent each constituent. For a discrete number of constituents (n), each having different affinities, the general sorption isotherm equation is reduced to

$$S = \sum_{n} F_i \Gamma(\zeta_i, C) \tag{1.7}$$

where F_i is the fraction relative to the total as discussed above. Conceptually, the adsorption site associated with each constituent provides an isotherm having its own affinity (ζ) and capacity. Consequently, a complete isotherm may be regarded as the sum of all individual isotherms. The function $g(\zeta)$ has been referred to as a "weighting function," a "site affinity distribution function," or a "frequency distribution of the affinity coefficient ξ for each constituent" (Limousin et al. 2007).

By choosing the accurate density function $g(\zeta)$, any type of isotherm can be described by the general isotherm equation (1.6; Hinz, Gaston, and Selim, 1994; Hinz, 2001). Consequently, various density functions that were capable of describing observed adsorption isotherms have been derived. Examples include Langmuir, two-site Langmuir, general Freundlich, general Langmuir-Freundlich, Freundlich, and Rothmund-Kkornfeld. For example, to describe Freundlich isotherms, Sposito (1980) suggested an affinity distribution function whose curve closely resembles a log-normal distribution.

As outlined by Kinniburgh et al. (1983) and later by Hinz, Gaston, and Selim (1994), a Langmuir isotherm can be derived by use of the Dirac delta function δ for $g(\zeta)$ as

$$g(\zeta) = \delta\,(\zeta - k) \tag{1.8}$$

Incorporation of Equation 1.8 in Equation 1.7 and integration yields the Langmuir isotherm,

$$\frac{S}{S_{max}} = \frac{kC}{1 + kC} \tag{1.9}$$

where k is an "overall" affinity coefficient that is equivalent to ω of Equation 1.3. Selecting $w(\zeta)$ in the form and proceeding as above yields the two-surface Langmuir isotherm equation,

$$S = S_{max}\left(\frac{F_1 k_1 C}{1 + k_1 C}\right) + \left(\frac{F_2 k_2 C}{1 + k_2 C}\right). \tag{1.10}$$

Consequently, a complete isotherm may be regarded as the sum of all individual Langmuir isotherms.

A major drawback of the above approach is that the distribution functions $g(\zeta)$ for the various isotherms are not very verifiable. In other words, there is no evidence or independent measure that verifies that such a specific distribution $g(\zeta)$ exists. In fact, distribution functions of the normal and log of normal type do not represent the constituents that make up the soil matrix. This is illustrated in Figure 1.2 for several commonly used probability distribution functions (PDFs). For most if not all soils, the dominant fractions are those associated with low-affinity sites such as sand and silt. In contrast, high-affinity sites, which are mainly associated with clay minerals, organic matter, and oxides, often are associated with much smaller fractions. Therefore, normal as well as log-normal distribution functions are unrealistic representations of the actual affinity distribution of a soil since it grossly underestimates the low-affinity fractions. Another drawback is that such affinity distribution functions $g(\zeta)$ are not considered unique. Therefore, distribution functions cannot be precluded from describing site affinity distributions.

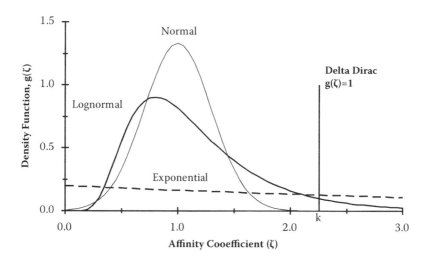

FIGURE 1.2
Probability distribution function for normal, log-normal, and exponential distributions and delta Dirac function.

In addition to the above shortcomings, another concern is the adoption of the Langmuir assumption for all solutes and soils. Although the Langmuir assumption is physically meaningful, it does not always describe isotherms for many solute species. In fact, for most if not all organic compounds, sorption isotherms cannot be described based on the Langmuir assumption. As an example, the atrazine isotherms for various smectites are shown in Figure 1.3 (Laird et al., 1992). These isotherms illustrate their linearity as well as extensive differences (i.e., heterogeneity) among the different smectites. Deviation from the Langmuir assumption, where maximum sorption is assumed, is likely due to multiple factors. It is possible that limited solubility of various trace elements and organic compounds in aqueous systems is one such factor.

In the following discussion, we deviate from the well-accepted Langmuir assumption as the governing sorption isotherm function and adopt the linear-type governing function. Specifically, we consider that for each constituent i of the soil system, sorption follows a simple linear form such that

$$\Gamma(\zeta_i, C) = \zeta_i C = k_i C \qquad where \qquad k_i = \zeta_i \qquad (1.11)$$

As a result, the isotherm equation for a heterogeneous system can be written as

$$S = \sum_{i=1}^{n} \chi(k_i, C) \qquad (1.12)$$

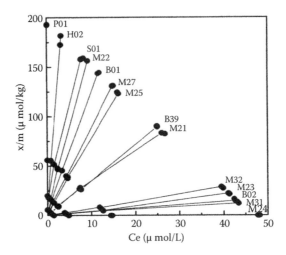

FIGURE 1.3
Isotherms for adsorption of atrazine on reference and smectites clay samples; atrazine adsorption was determined at four levels of the Ca-saturated clay samples in 0.01 M CaCl$_2$. H02, hectrite, B01, B02, M21, M22, M23, M25, M27, M32, B39, montomorillonite; M31, smectites/illite; S01, saponite; P01, beidellite. (From Laird et al., 1992. With permission.)

where n is the number of soil fractions or constituents and

$$\chi(k_i, C) = \begin{cases} k_i C & C \le \sigma_i \\ S_{max} F_i & C \ge \sigma_i \end{cases} \tag{1.13}$$

$$\sigma_i = S_{max}(F_i/k_i) \tag{1.14}$$

and $\chi(k_i, C)$ is a piece-wise continuous function for each constituent i. This function is necessary in order to comply with the constraint that maximum sorption cannot be exceeded, in accordance with Equation 1.4. The above expressions implies that for a given constituent i, linear sorption takes place (C versus S). As C increases S increases linearly when all fractions participate. Once the sorption capacity for the strongest fraction F_n is reached, no additional sorption for such a fraction can take place. This will hold true with increasing C until maximum sorption for the subsequent fraction, in this case the fraction F_{n-1}, is reached, and so on.

A consequence of the above formulation is that at all concentrations where the sorption maximum was not attained for any one constituent; we have a simple linear sorption isotherm for a heterogeneous system:

$$S = \lambda C, \tag{1.15}$$

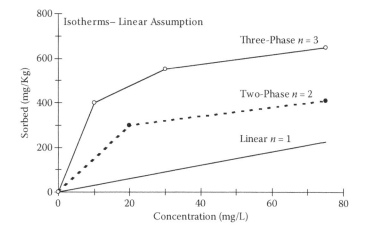

FIGURE 1.4
Linear, two-, and three-phase isotherms based on the linear model assumption.

where

$$\lambda = \sum_{i=1}^{n} F_i k_i \tag{1.16}$$

This equation resembles that of $S = K_d C$, where λ is a weighted average of the affinity coefficient of all constituents. Based on the above we present, in Figure 1.4, three isotherms: linear, two-, and three-phase isotherms, where two ($n = 2$) and three ($n = 3$) constituents were assumed. There are numerous examples in the literature of two- and three-phase isotherms.

Extending the analysis to a soil matrix having ten constituents results in the nonlinear isotherms shown in Figure 1.5. In both isotherms, ten equal fractions were assumed, where the affinity varied linearly from 30 to 100 for isotherm A and 50 to 100 for isotherm B. The isotherms illustrate nonlinear Freundlich-type behavior with the exponent parameter $b < 1$; $b = 0.59$ and 0.46 for isotherms A and B, respectively. A complete isotherm with a greater number of constituents may be regarded as the sum of all individual isotherms in a similar manner to that for Langmuir isotherms described above. The only exception is that all isotherms are subject to the constraint given by the piece-wise continuous function, Equation 1.13. Based on these results, Freundlich behavior can be described without the constraint of the log-normal distribution of site affinities.

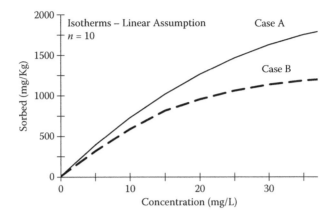

FIGURE 1.5
Nonlinear isotherms with 10 fractions ($n = 1$) based on the linear model assumption.

1.6 Effect of Soil Properties

Buchter et al. (1989) studied the retention of 15 elements by 11 soils from 10 soil orders to determine the effects of element and soil properties on the magnitude of the Freundlich parameters K_f and b. They also explored the correlation of the Freundlich parameters with selected soil properties and found that pH, cation-exchange capacity (CEC), and iron/aluminum oxide content were the most important factors for correlation with the partitioning coefficients. The names, taxonomic classification, and selected properties of the 11 soils used in their study are listed in Table 1.2, and estimated values for K_f and b for selected heavy metals are given in Table 1.3. A wide range of K_f values, from 0.0419 to 4.32×10^7 ml g^{-1}, were obtained, which illustrates the extent of heavy metals affinity among various soil types. Such a wide range of values was not obtained for the exponent parameter b, however. The magnitude of K_f and b was related to both soil and element properties. Strongly retained elements such as Cu, Hg, Pb, and V had the highest K_f values. The transition metal cations Co and Ni had similar K_f and b values, as did the group IIB elements Zn and Cd. Oxyanion species tended to have lower b values than did cation species. Soil pH and CEC were significantly correlated to log K_f values for cation species. High pH and high CEC soils retained greater quantities of the cation species than did low pH and low CEC soils. A significant negative correlation between soil pH and the Freundlich parameter b was observed for cation species, whereas a significant positive correlation between soil pH and b for Cr(VI) was found. Greater quantities of anion species were retained by soils with high amounts of amorphous iron oxides, aluminum oxides, and amorphous material than were retained by soils with

TABLE 1.2

Freundlich Model Parameters K_f and b for Selected Soils and Heavy Metals

Element	Initial Species	Soil											
		Alligator	Calciorthid	Cecil	Kula	Lafitte	Molakai	Norwood	Olivier	Oldsmar	Webster	Windsor	Mean*
		K_f (mL/g)											
Co	Co^{2+}	3.57E+01	2.51E+02	6.56E+00	1.05E+02	3.39E+01	9.25E+01	2.74E+01	6.70E+01	2.55E+00	3.63E+02	6.28E+00	3.75E+01
Ni	Ni^{2+}	3.78E+01	2.06E+02	6.84E+00	1.10E+02	5.01E+01	4.49E+01	2.09E+01	5.05E+01	3.44E+00	3.37E+02	8.43E+00	3.61E+01
Cu	Cu^{2+}	2.58E+02	2.62E+03	5.37E+01	2.05E+03	2.21E+02	3.68E+02	8.91E+01	2.18E+02	5.62E+01	6.35E+03	7.71E+01	3.17E+03
Zn	Zn^{2+}	2.81E+01	4.20E+02	1.12E+01	2.38E+02	2.01E+01	8.04E+01	4.21E+01	8.91E+01	2.12E+00	7.74E+02	9.68E+00	4.79E+01
Cd	Cd^{2+}	5.25E+01	2.88E+02	1.39E+01	1.87E+02	5.27E+01	9.12E+01	2.88E+01	9.79E+01	5.47E+00	7.55E+02	1.44E+01	5.93E+01
Hg	Hg^{2+}	1.08E+02	1.96E+01	8.13E+01	2.49E+02	1.90E+02	1.20E+02	1.13E+02	1.29E+02	8.63E+01	2.99E+02	1.30E+02	1.15E+02
Pb	Pb^{2+}	1.81E+03		2.36E+02	4.32E+07	9.18E+02	8.17E+03	3.85E+02	1.64E+04	1.36E+02		4.72E+02	3.37E+02
V	VO_3^-	1.42E+02	1.08E+01	3.97E+01	2.22E-03	1.03E+02	5.05E+02	1.86E+01	9.12E+01	9.08E+01	8.07E+01	1.53E+01	1.03E+02
Cr	CrO_4^{2-}	3.41E+00			6.28E+01	3.03E+01	6.41E+00			5.47E+00		8.47E+00	1.12E+01
Mo	$Mo_7O_{24}^{6-}$	5.75E+01		1.80E+01	4.11E+02	8.15E+01	1.18E+02			2.56E+01		4.38E+01	6.44E+01
As	AsO_4^{3-}	4.78E+01	8.87E-00	1.98E+01	1.50E+03	7.10E+01	1.56E-02	8.53E+00	4.60E+01	1.87E+01	2.36E+01	1.05E+02	4.71E+01
		b											
Co	Co^{2+}	0.953	0.546	0.745	0.878	1.009	0.621	0.627	0.584	0.811	0.782	0.741	0.754
Ni	Ni^{2+}	0.939	0.504	0.688	0.738	0.903	0.720	0.661	0.646	0.836	0.748	0.741	0.739
Cu	Cu^{2+}	0.544	1.140	0.546	1.016	0.987	0.516	0.471	0.495	0.602	1.420	0.567	0.755
Zn	Zn^{2+}	1.011	0.510	0.724	0.724	0.891	0.675	0.515	0.625	0.962	0.697	0.792	0.739
Cd	Cd^{2+}	0.902	0.568	0.768	0.721	0.850	0.773	0.668	0.658	0.840	0.569	0.782	0.736
Hg	Hg^{2+}	0.741	0.313	0.564	1.700	0.751	0.960	0.582	1.122	0.513	2.158	0.681	1.008
Pb	Pb^{2+}	0.853		0.662	5.385	0.558	1.678	0.741	0.998	0.743		0.743	1.485
V	VO_3^-	0.592	0.857	0.629	1.402	0.679	0.847	0.877	0.607	0.483	0.762	0.647	0.762
Cr	CrO_4^{2-}	0.504			0.609	0.374	0.607			0.394		0.521	0.501
Mo	$Mo_7O_{24}^{6-}$	0.882		0.617	1.031	0.607	0.664			0.451		0.544	0.685
As	AsO_4^{3-}	0.636	0.554	0.618	1.462	0.747	0.561	0.510	0.548	0.797	0.648	0.601	0.698

TABLE 1.3

Taxonomic Classification and Selected Chemical and Physical Properties of the Soils Named in Table 1.2

Soil[a]	Horizon	Taxonomic Classification	pH	TOC %	Sum of Cations CEC cmol/kg	Exch. OH	MnO₂ %	Amor. Fe_2O_3	Free Fe_2O_3 %	Al_2O_3 %	$CaCO_3$ %	Sand	Silt	Clay
Alligator	Ap	Very fine, morillonitic, acid, thermic Vertic Haplaquept	4.8	1.54	30.2	3.5	0.028	0.33	0.74	0.15	—	5.9	39.4	54.7
Unnamed	Ap	Calciorthid	8.5	0.44	14.7	33.8	0.015	0.050	0.25	0.000	7.39	70.0	19.3	10.7
Cecil	Ap	Clayey, kaolinitic, thermic Typic Hapludult	5.7	0.61	2.0	0.011	0.099	1.76	0.27	—	67.7	12.8	7.3	
Cecil	B	Clayey, kaolinitic, thermic Typic Hapludult	5.4	0.26	2.4	6.6	0.002	0.082	7.48	0.94	—	30.0	18.8	51.2
Kula	Ap1	Medial, isothermic Typic Euthandept	5.9	6.62	22.5	82.4	0.093	1.68	5.85	3.51	—	73.7	25.4	0.9
Kula	Ap2	Medial, isothermic Typic Euthandept	6.2	6.98	27.0	58.5	0.13	1.64	6.95	3.67	—	66.6	32.9	0.5
Lafitte	Ap	Euic, thermic Typic Medisaprist	3.9	11.6	26.9	4.7	0.009	1.19	1.16	0.28	—	60.7	21.7	17.6
Molokai	Ap	Clayey, kaolinitic, isohyperthermic Typic Torrox	6.0	1.67	11.0	7.2	0.76	0.19	12.4	0.91	—	25.7	46.2	28.2
Norwood	Ap	Fine-silty, mixed (calc.), thermic Typic Udifluvent	6.9	0.21	4.1	0.0	0.008	0.061	0.30	0.016	—	79.2	18.1	2.8
Olivier	Ap	Fine-silty, mixed,	6.6	0.83	8.6	1.9	0.27	0.30	0.71	0.071	—	4.4	89.4	6.2

		thermic Aquic Fragiudalf										
Unnamed	B21h	Spodosol	4.3	1.98	2.7	5.2	0.0000	0.009	0.008	0.22	—	90.2	6.0	3.8
Webster	Ap	Fine-loamy, mixed, mesic Typic Haplaquol	7.6	4.39	48.1	14.1	0.063	0.19	0.55	0.10	3.14	27.5	48.6	23.9
Windsor	Ap	Mixed, mesic Typic Udipsamment	5.3	2.03	2.0	10.2	0.041	0.42	1.23	0.56	—	76.8	20.5	2.8
Windsor	B	Mixed, mesic	5.8	0.67	0.8	10.1	0.031	0.23	0.79	0.29	—	74.8	24.1	1.1

[a] The states from which the soil samples originated are Louisiana (Alligator, Lafitte, Norwood, and Olivier soils), South Carolina (Cecil soil), Hawaii (Kula and Molokai soils), Iowa (Webster soil), New Hampshire (Windsor soil), New Mexico (Calciorthid), and Florida (Spodosol).

low amounts of these minerals. Several anion species were not retained by high pH soils. Despite the fact that element retention by soils is the result of many interacting processes and that many factors influence retention, significant relationships among retention parameters and soil and element properties exist even among soils with greatly different characteristics. Buchter et al. (1989) made the following conclusions:

1. pH is the most important soil property that affects K_f and b.
2. CEC influences K_f for cation species.
3. The amounts of amorphous iron oxides, aluminum oxides, and amorphous material in soils influence both cation and anion retention parameters.
4. Except for Cu and Hg, transition metal (Co and Ni) and group IIB cations (Zn and Cd) have similar K_f and b values for a given soil.
5. Significant relationships between soil properties and retention parameters exist even in a group of soils with greatly different characteristics.

The relationships between soil properties and retention parameters (e.g., Figure 1.6) can be used to estimate retention parameters when retention data for a particular element and soil type are lacking, but soil property data are available. For example, the retention characteristics of Co, Ni, Zn, and Cd are sufficiently similar that these elements can be grouped together

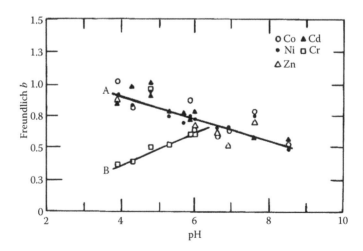

FIGURE 1.6
Correlation between soil pH and Freundlich parameter b. Curve A is a regression line for Co, Ni, Zn, and Cd ($b = 1.24 - 0.0831$ pH, r = 0.83"). Curve B is for Cr(VI) ($b = -0.0846 + 0.116$ pH, r = 0.98"). (After Buchter et al., 1989.)

and an estimated b value for any one of them could be estimated from soil pH data using the regression equation for curve A in Figure 1.2. For many purposes such an estimate would be useful, at least as a first approximation, in describing the retention characteristics of a soil.

$$\text{Curve A:} \quad b = 1.24 - 0.0831 \text{ pH} \quad (r = 0.83) \quad (1.17)$$

$$\text{Curve B:} \quad b = -0.0846 + 0.116 \text{ pH} \quad (r = 0.98) \quad (1.18)$$

Sauvé, Hendershot, and Allen (2000) analyzed more than 70 studies of various origins collected from the literature in an effort to correlate the distribution coefficient K_d with soil properties for five heavy metals, cadmium, copper, lead, nickel, and zinc. Specifically, the relationships between the reported K_d values were explored relative to variations in soil solution pH, soil organic carbon (SOC), and total metal retained by the soil. Sauvé, Hendershot, and Allen (2000) proposed two models to predict K_d values for several heavy metals based on chemical properties of the soil. These models were developed based on regression analysis of extensive K_d values published in the literature for a wide range of soils. The proposed models are

Model I $$\log(K_d) = \mathbf{A} + \mathbf{B} \text{ pH} \quad (1.19)$$

Model II $$\log(K_d) = \mathbf{A} + \mathbf{B} \text{ (pH)} + \mathbf{C} \text{ (SOC)} \quad (1.20)$$

where A, B, and C are fitting parameters. Sauvé, Hendershot, and Allen (2000) proposed a third model, which incorporated retained heavy metals by the soil prior to K_d measurements. With the exception of contaminated sites, the amounts of heavy metals retained are often extremely small, however. In Figure 1.7, plots of K_d values on a log scale versus soil pH are presented for all five elements based on a compilation by Sauvé, Hendershot, and Allen (2000).

In 2013, Selim and coworkers (unpublished data) studied the retention of five heavy elements by 10 soils from 10 soil orders to determine the effects of element and soil properties on the magnitude of the Freundlich parameters K_f and b. They measured adsorption after one and seven days, as well as desorption following adsorption (based on successive dilution method; see Chapter 2). The metals investigated were cadmium, copper, lead, nickel, and zinc. These five metals are the same considered by Sauvé, Hendershot, and Allen (2000) in their regression study. The names, taxonomic classifications, and selected properties of the 10 soils used in this study are listed in Table 1.4. The Ap horizons of all soils were used in this retention study. The only exception is that the sandy candor subsurface sample was sampled at a depth of 90 cm. Physical and chemical properties of the 10 soils were also quantified. A comparison of K_d parameter values for all 10 soils and five heavy metals is provided in Table 1.5 for sorption at one and seven days. Freundlich model K_f and b parameter values for all 10 soils and five heavy metals are provided in

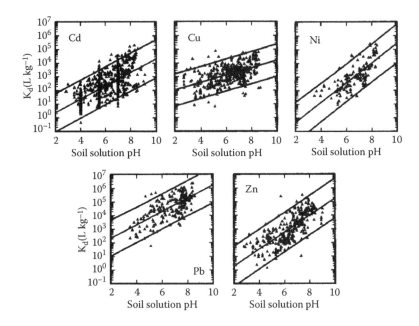

FIGURE 1.7
Plots of K_d versus soil pH for Cd, Cu, Ni, Pb, and Zn based on a compilation by Sauvé, Hendershot, and Allen (2000). The upper and lower lines represent the 95% confidence intervals. (From Sauvé, Hendershot, and Allen, 2000. With permission.)

Table 1.9 for sorption at one and seven days. These results illustrate the sorption affinity of the various soils for each of the heavy metals.

Copper isotherms for Houston and Olney soils are shown in Figures 1.8 and 1.9. The results indicate extremely high sorption of copper by both soils as manifested by the low concentration in the soil solution after 24 hours of sorption. This high sorption is due to the high clay content of Houston clay as well as the presence of carbonates. The dashed and solid curves represent linear and Freundlich model simulations. For the Candor sand soils shown in Figure 1.9, the isotherms indicate the lowest copper sorption. In contrast, the Arapahoe soil shown in Figure 1.8 indicates extensive copper sorption. Solid curves are simulations using the Freundlich model, which best described the isotherms for all 10 soils. Best-fit parameter values for the linear and Freundlich models along with their r^2 are given in Table 1.3.

Cadmium isotherms for all soils are shown in Figures 1.10 and 1.11. Houston, Arapahoe, and Sharkey soils exhibited the highest cadmium sorption indicative of strong copper affinity for soils with high clay content and organic matter. The lowest sorption for copper is illustrated in

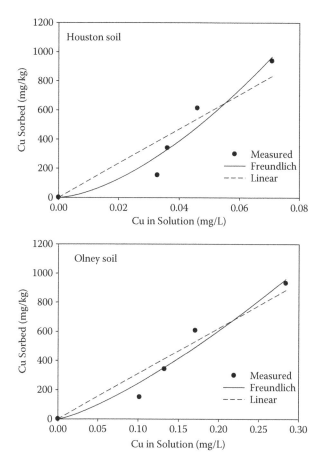

FIGURE 1.8
Copper isotherms for Houston and Olney soils after 1 day of reaction. Solid and dashed curves are simulations using the linear and Freundlich models.

Figure 1.10 for Lincoln, Nada, and Candor soils. Solid and dashed curves in Figures 1.6 and 1.7 are simulations using the Freundlich model, which best described the isotherms for all 10 soils. Parameter values for K_d of the linear model and K_f and b for the Freundlich models along with their r^2 are given in Table 1.4.

Zinc isotherms for all soils are shown in Figures 1.12 and 1.13. These results indicate much less affinity for zinc by all 10 soils when compared with their affinity for copper. Houston soil exhibited the highest sorption, whereas most loamy soils indicated moderate sorption for zinc. For Candor

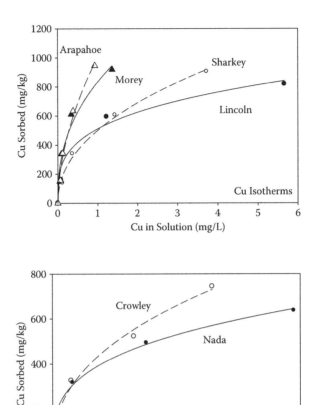

FIGURE 1.9
Copper isotherms for several soils after one day of reaction. Solid curves are simulations using the Freundlich models.

sand soils low sorption for zinc is shown where the shape of the isotherms was the opposite of those for all other soils. The extent of the nonlinearity and shape of the isotherms is illustrated by the b values of the Freundlich equation given in Table 1.5. Values for b are less than 1 for all soils, except for candor sand soils where b was greater than 1, indicative of irreversible reactions. For linear isotherms, the parameter b is 1. This parameter is often regarded as a measure of the extent of the heterogeneity of sorption sites on the soil having different affinities for solute retention by matrix surfaces.

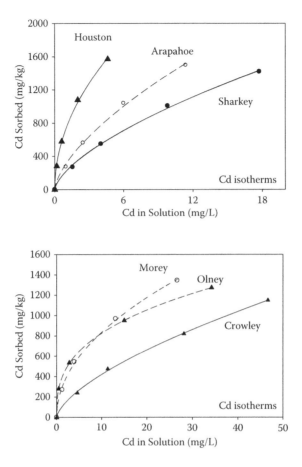

FIGURE 1.10
Cadmium isotherms for several soils after one day of reaction. Solid and dashed curves are simulations using the Freundlich model.

In a heterogeneous system, sorption by the highest energy sites takes place preferentially at the lowest solution concentrations, and as the sorbed concentration increases, successively lower energy sites become occupied. This leads to a concentration-dependent sorption equilibrium behavior, that is, a nonlinear isotherm.

Nickel isotherms for all soils are shown in Figures 1.14 and 1.15. These results indicate similar affinities for nickel by all 10 soils when compared to zinc. Houston soil exhibited the highest sorption, whereas most loamy soils indicated moderate sorption affinities. Among all soils, Candor sand soils exhibited the lowest affinity for nickel. Parameter values for K_d of the linear

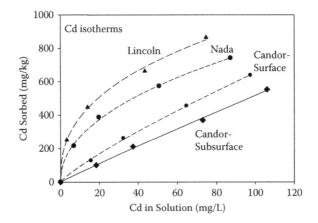

FIGURE 1.11
Cadmium isotherms for four soils after one day of reaction. Solid and dashed curves are simulations using the Freundlich model.

model and K_f and b for the Freundlich models along with their r^2 are given in Table 1.6.

Lead isotherms for all soils are shown in Figures 1.16 and 1.17. These results indicate a wide range of affinities for lead in the 10 soils. Houston soil exhibited the highest sorption where the solution of lead in the soil solution was below detection for the entire range of input concentration.

TABLE 1.4

Selected Soil Properties of the 10 Soils Used in the Study

Soil Series	State	Texture	pH	%C	CEC	Sand %	Silt %	Clay %	Carbonates (%)
Arapahoe	NC	FSL	5.02	10.22	30.10	64.7	25.2	10.1	
Candor–Surface	NC	LCoS	4.39	2.24	5.30	84.1	6.6	9.3	
Candor–Subsurface	NC	CoS	4.05	0.56	1.70	91.0	4.1	4.9	
Olney	CO	FSL	8.12	1.14	10.10	71.1	11.6	17.3	3.32
Lincoln	OK	VFSL	7.54	1.27	3.00	59.5	29.7	10.8	3.6
Nada	TX	FSL	6.61	0.76	6.30	56.8	33.6	9.6	
Morey	TX	L	7.74	1.05	27.80	29.4	46.5	24.1	
Crowley	LA	SiL	5.22	1.16	16.50	8.3	77.3	14.4	
Sharkey	LA	SiC	5.49	2.76	39.70	6.4	48.6	45.0	
Houston	TX	SiC	7.78	4.26	47.70	9.6	41.2	49.2	26.0

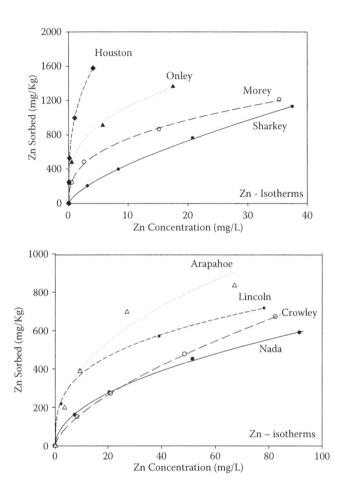

FIGURE 1.12

Zinc isotherms for several soils after one day of reaction. Solid and dashed curves are simulations using the Freundlich model.

The detection limit for lead using inductively coupled plasma-atomic emission spectrometry (ICP-AES) was 28 µg/L (ppb). The other soils exhibited different degrees of affinity for lead, as illustrated in Figures 1.16 and 1.17. Soils with high organic matter content such as Arapahoe as well as soils with high clay contents such as Sharkey exhibited high sorption. It should also be emphasized that Olney soil with high pH and carbonate content exhibited high sorption for lead. Parameter values for K_d of the linear model and K_f and b for the Freundlich models along with their r^2 are given in Table 1.7.

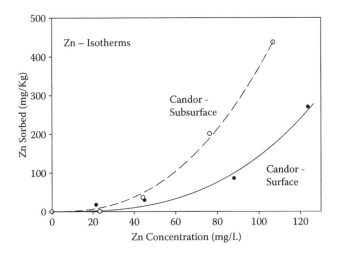

FIGURE 1.13
Zinc isotherms for Candor soil after one day of reaction. Curves are simulations using the
Freundlich models.

TABLE 1.5

Taxonomic Classification of the Ten Soils Used in the Study

Soil Series	State	Taxonomic Classification
Arapahoe	NC	Coarse-loamy, mixed, semiactive, nonacid, thermic Typic Humaquepts
Candor–Surface	NC	Sandy, kaolinitic, thermic Grossarenic Kandiudults
Candor–Subsurface	NC	Sandy, kaolinitic, thermic Grossarenic Kandiudults
Olney	CO	Fine-loamy, mixed, superactive, mesic Ustic Haplargids
Lincoln	OK	Sandy, mixed, thermic Typic Ustifluvents
Nada	TX	Fine-loamy, siliceous, active, hyperthermic Albaquic Hapludalfs
Morey	TX	Fine-silty, siliceous, superactive, hyperthermic Oxyaquic Argiudolls
Crowley	LA	Fine, smectitic, thermic Typic Albaqualfs
Sharkey	LA	Fine-silty, mixed, active, thermic Typic Glossaqualfs
Houston	TX	Fine, smectitic, thermic Udic Haplusterts

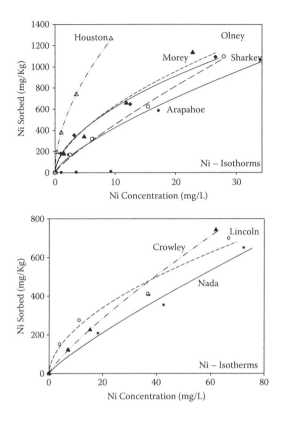

FIGURE 1.14
Nickel isotherms for several soils after one day of reaction. Solid and dashed curves are simulations using the Freundlich model.

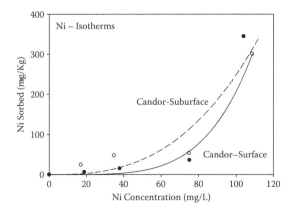

FIGURE 1.15
Nickel isotherms for Candor surface and subsurface soils after one day of reaction. Solid and dashed curves are simulations using the Freundlich model.

TABLE 1.6
Linear Model K_d Parameter Values for 10 Soils and Five Heavy Metals

Element	Arapahoe	Candor–Surface	Candor–Subsurface	Olney	Lincoln	Nada	Morey	Crowley	Sharkey	Houston
K_d 1 day										
Cu	1142.28	6.23	2.52	3110.95	161.45	25.31	754.59	42.63	272.98	11748.66
Zn	16.51	1.66	3.21	86.30	10.65	7.30	38.45	8.83	32.32	415.76
Cd	144.68	6.82	5.21	42.34	13.09	9.69	56.67	26.68	87.84	380.02
Ni	32.47	1.98	2.10	43.98	10.96	8.93	47.92	11.91	39.99	147.05
Pb	4035.40	19.92	5.46	34996.46	67.83	41.65	20696.37	109.84	1344.13	ND**
K_d 7 days										
Cu	1776.43	11.49	5.59	4680.96	3406.10	32.13	5233.00	53.11	420.37	13124.76
Zn	17.91	2.26	2.94	384.38	19.83	8.00	47.32	9.26	33.60	817.07
Cd	156.42	7.02	5.51	60.85	15.57	10.62	59.16	27.47	87.62	664.69
Ni	38.68	1.95	2.19	111.64	34.28	9.19	44.51	12.23	42.70	281.76
Pb	5497.26	26.47	7.67	57251.48	4329.19	53.63	85573.42	121.94	2311.93	ND

a　Not detected.
** Nd = Not detected.

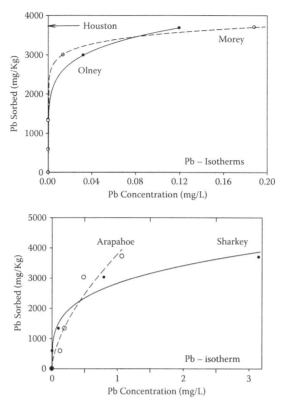

FIGURE 1.16
Lead isotherms for several soils after one day of reaction. Curves are simulations using the Freundlich models.

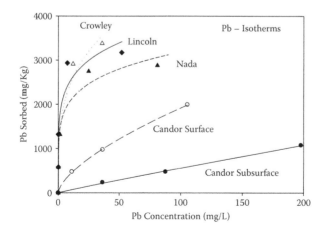

FIGURE 1.17
Lead isotherms for five different soils after one day of reaction. Curves are simulations using the Freundlich models.

TABLE 1.7

Freundlich Model K_f and b Parameter Values for Ten Soils and Five Heavy Metals

Element	Arapahoe	Candor-Surface	Candor-Subsurface	Olney	Lincoln	Nada	Morey	Crowley	Sharkey	Houston
K_f 1 day										
Cu	1004.91	64.51	16.15	5027.78	512.40	235.45	833.78	213.88	501.73	66205.54
Zn	150.51	0.00	0.00	552.83	181.27	60.29	323.29	37.21	91.07	988.81
Cd	313.18	13.62	6.30	367.21	157.04	92.94	267.61	93.42	220.88	738.24
Ni	78.51	–	0.00	169.79	64.04	16.65	159.24	17.72	65.92	373.10
Pb	3792.51	93.99	7.09	5169.46	1679.58	1360.44	4224.09	1280.45	2825.04	C S
K_f 7 days										
Cu	1305.73	103.12	72.88	2209492.85	3993.56	350.52	4842.30	283.91	630.16	29606459.71
Zn	22.37	0.51	1.53	847.56	243.53	50.98	387.64	45.00	96.24	1258.36
Cd	323.09	16.68	7.94	514.97	0.39	93.58	301.51	103.19	224.12	935.35
Ni	89.39	–	0.00	261.10	134.28	28.04	130.48	23.51	70.94	467.14
Pb	4705.95	174.11	33.53	6889.88	3863.83	1497.59	5296.20	1452.22	3275.69	C S
B, 1 day										
Cu	0.54	0.40	0.54	1.32	0.29	0.30	0.41	0.41	0.46	1.16
Zn	0.43	2.85	2.50	0.31	0.32	0.51	0.37	0.66	0.70	0.33
Cd	0.65	0.84	0.96	0.35	1.00	0.47	0.49	0.65	0.65	0.50

Ni	0.73	–	2.74	0.56	0.56	0.85	0.60	0.90	0.84	0.54
Pb	0.62	0.65	0.95	0.16	0.18	0.18	0.08	0.28	0.27	C S
b, 7 days										
Cu	0.58	0.39	0.34	3.96	1.10	0.22	0.96	0.35	0.44	3.48
Zn	0.83	1.32	1.14	0.37	0.33	0.56	0.34	0.62	0.69	0.28
Cd	0.66	0.80	0.92	0.30	0.33	0.48	0.46	0.63	0.65	0.54
Ni	0.74	–	2.83	0.62	0.57	0.73	0.65	0.83	0.84	0.64
Pb	0.69	0.56	0.70	0.24	0.36	0.18	0.12	0.26	0.29	C S

1.7 Predictions

The models proposed by Sauvé, Hendershot, and Allen (2000) to predict K_d values given in Equations 1.3 and 1.4 were tested for the ability to predict measured K_d values for the five heavy metals and 10 soils discussed above. As discussed earlier, for model I, K_d predictions are based on soil pH only. In model II, K_d predictions were based on two variables, pH and percent soil organic carbon (SOC). Comparison of measured K_d values from this study for all 10 soils using models I and II are shown in Figures 1.18 to 1.22. For copper, predictions were overestimated at the low K_d range. For both cadmium and zinc, the predictions did not illustrate any pattern and were highly inadequate. Somewhat improved predictions were obtained for nickel and lead when model I was used. Overall, both models yielded inadequate predictions for all heavy metals used in this study. In contrast, when the Buchter et al. (1989) model was used to predict the Freundlich parameter b, good overall predictions were obtained (see Figure 1.23). For all heavy metals extremely good trends were observed. The best prediction was obtained for Zn. Based on this investigation, the following conclusions can be drawn.

1 Adsorption of all five heavy metals was nonlinear.
2 For all 10 soils used in this study, adsorption of heavy metals follows the order Pb > Cu > Cd > Zn > Ni. In the presence of carbonates, adsorption of heavy metals follows the order Pb > Cu > Zn > Cd > Ni.

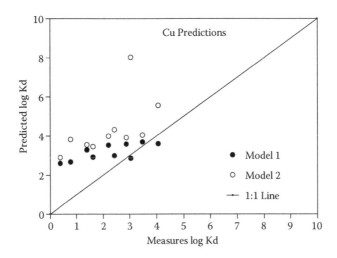

FIGURE 1.18
Measured and calculated K_d values for Cu for all soils. Calculated values for K_d were obtained using models 1 and 2 of Sauvé, Hendershot, and Allen (2000).

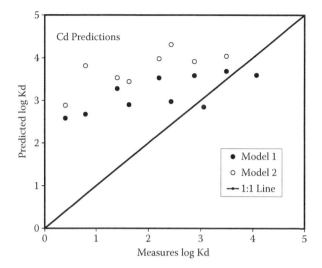

FIGURE 1.19
Measured and calculated K_d values for Cd for all soils. Calculated values for K_d were obtained using models 1 and 2 of Sauvé, Hendershot, and Allen (2000).

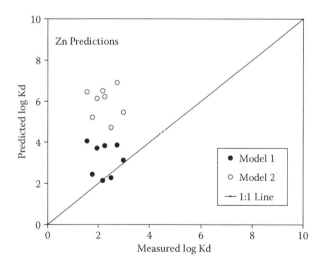

FIGURE 1.20
Measured and calculated K_d values for Zn for all soils. Calculated values for K_d were obtained using models 1 and 2 of Sauvé, Hendershot, and Allen (2000).

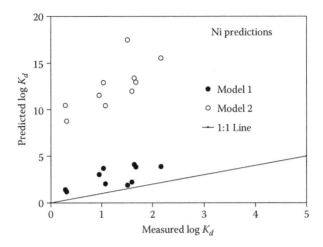

FIGURE 1.21
Measured and calculated K_d values for Ni for all soils. Calculated values for K_d were obtained using models 1 and 2 of Sauvé, Hendershot, and Allen (2000).

For all 10 soils used in this study, the Sauvé, Hendershot, and Allen (2000) models based on soil pH or soil pH and organic carbon provided less than adequate predictions for K_d values for all five heavy metals. In contrast, good predictions were obtained for the Freundlich parameter b when the Buchter et al. (1989) model was used.

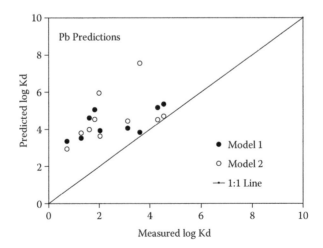

FIGURE 1.22
Measured and calculated K_d values for Pb for all soils. Calculated values for K_d were obtained using models 1 and 2 of Sauvé, Hendershot, and Allen (2000).

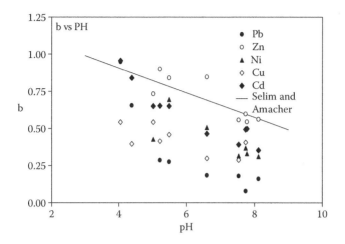

FIGURE 1.23

Measured and calculated Freundlich *b* parameter values for all heavy metals and soils. Calculated values for *b* were obtained using Buchter et al. (1989).

References

Bradl, H. B. 2004. Adsorption of Heavy Metal Ions on Soils and Soil Constituents. *J. Colloid Interface Sci.* 277: 1–18.

Buchter, B., B. Davidoff, M. C. Amacher, C. Hinz, I. K. Iskandar, and H. M. Selim. 1989. Correlation of Freundlich K_d and n Retention Parameters with Soils and Elements. *Soil Sci.* 148: 370–379.

Environmental Protection Agency. 1999. Understanding Variation in Partition Coefficient, Kd, Values. Volume I: The Kd Model, Methods of Measurement, and Application of Chemical Reaction Codes. EPA 402-r-99-004a (212p).

Fontes, M. 2012. Behavior of Heavy Metals in Soils: Individual and Multiple Competitive Adsorption. In *Competitive Sorption and Transport of Trace Elements in Soils and Geological Media*, edited by H. M. Selim, 77–117. Boca Raton, FL: CRC Press.

Fuller, C. C., J. A. Davis, and G. A. Waychunas. 1993. Surface Chemistry of Ferrihydrite: Part 2. Kinetics of Arsenate Adsorption and Coprecipitation. *Geochim. Cosmochim. Acta* 57: 2271–2282.

Giles, C. H., A. P. D'Silva, and I. A. J. Easton. 1974. A General Classification of the Solute Adsorption Isotherms II. *J. Colloid Interface Sci.* 47: 766–778.

Goldberg, S. 1992. Use of Surface Complexation Models in Soil Chemical Systems. *Adv. Agron.* 47: 233–329.

Hinz, C., 2001. Description of Sorption Data with Isotherm Equations. *Geoderma* 99: 225–243.

Hinz, C., L. A. Gaston, and H. M. Selim. 1994. Effect of Sorption Isotherm Type on Predictions of Solute Mobility in Soil. *Water Resour. Res.* 30: 3013–3021.

Holford, I. C. R., and G. E. G. Mattingly. 1975. The High- and Low-Energy Phosphate Adsorption Surfaces in Calcareous Soils. *J. Soil Sci.* 26: 407–417.

Holford, I. C. R., R. W. M. Wedderburn, and G. E. G. Mattingly. 1974. A Langmuir Two-Surface Equation as a Model of Phosphate Adsorption by Soils. *J. Soil Sci.* 25: 242–254.

Kinniburgh, D. G. 1986. General Purpose Adsorption Isotherms. *Environ. Sci. Technol.* 20, 895–904.

Kinniburgh, D.G., Barker, J.A., Whitefield, M., 1983. A Comparison of Some Simple Adsorption Isotherms for Describing Divalent Cation Adsorption by Ferrihydrite. *J. Colloid Interf. Sci.* 95:370–384.

Laird, D.A., E. Barriuso, R.H. Dowdy, and W.C. Koskinen. 1992. Adsorption of Atrazine on Smectites. *Soil Sci. Soc. Am. J.* 56:62-67.

Langmuir, I. 1918. The Adsorption of Gases on Plane Surfaces of Glass, Mica and Platinum. *J. Am. Chem. Soc.* 40: 1361–1402.

Limousin, G., J.-P. Gaudet, L. Charlet, S. Szenknect, V. Barthes, and M. Krimissa. 2007. Sorption Isotherms: A Review on Physical Bases, Modeling and Measurement. *Appl. Geochem.* 22: 249–275.

Raven, K. P., A. Jain, and R. H. Loeppert. 1998. Arsenite and Arsenate Adsorption on Ferrihydrite: Kinetics, Equilibrium, and Adsorption Envelopes. *Environ. Sci. Technol.* 32: 344–349.

Sauvé, S., W. Hendershot, and H. Allen. 2000. Solid-Solution Partitioning of Metals in Contaminated Soils: Dependence on pH, Total Metal Burden, and organic Matter. *Environ. Sci. Technol.* 34: 1125–1131.

Schmidt, H. W., and H. Sticher. 1986. Long-Term Trend Analysis of Heavy Metal Content and Translocation in Soils. *Geoderma* 38: 195–207.

Sparks, D. L. 2003. *Environmental Soil Chemistry,* second edition. New York: Academic Press.

Sposito, G. 1980. Derivation of the Freundlich Equation for Ion Exchange Reactions in Soils. *Soil Sci. Soc. Am. J.* 44: 652.

Zhang, H. 2013. Modeling Surface Complexation Kinetics. ICOBTE 2013 Proceedings, Athens, GA 474–475.

2

Sorption Kinetics

For several decades, it has been observed that sorption and desorption of various chemicals on matrix surfaces are kinetic or time dependent. Numerous studies on the kinetic behavior of solutes in soils are available in the literature. Recent reviews on kinetics include Sparks and Suarez (1991), Sparks (2003), and Carrillo-Gonzalez et al. (2006). The extent of kinetics varied extensively among the different solute species and soils considered. Generally, trace elements and heavy metal species exhibit strong sorption, as well as extensive kinetic behavior during sorption and release or desorption. In contrast, weak sorption and less extensive kinetic behavior are often observed for organic chemicals in soils and porous media. According to Aharoni and Sparks (1991) and Sparks (2003), a number of transport and chemical reaction processes affect the rate of soil chemical reactions. The slowest of these will limit the rate of a particular reaction. The actual chemical reaction at the surface, for example, adsorption, is usually very rapid and not rate limiting. Transport processes (see Figure 2.1) include (1) transport in the solution phase, which is rapid and, in the laboratory, can be eliminated by rapid mixing; (2) transport across a liquid film at the particle/liquid interface (film diffusion); (3) transport in liquid-filled macropores (>2 nm), all of which are nonactivated diffusion processes and occur in mobile regions; (4) diffusion of a sorbate along pore wall surfaces (surface diffusion); (5) diffusion of sorbate occluded in micropores (<2 nm—pore diffusion); and (6) diffusion processes in the bulk of the solid, all of which are activated diffusion processes. Pore and surface diffusion can be referred to as interparticle diffusion, whereas diffusion in the solid is intraparticle diffusion.

The form of chemical retention reactions in soils and geological porous media must be clearly identified if predictions of their potential mobility, toxicity, and impact on the environment are sought. In general, chemical retention processes with matrix surfaces have been quantified by scientists using a number of empirically based approaches. One approach represents equilibrium-type reactions such as those discussed in Chapter 1.1 Equilibrium models are those where sorption reactions are assumed fast or instantaneous in nature. Under such conditions, "apparent equilibrium" may be observed in a relatively short reaction time (minutes or hours). Langmuir and Freundlich models are perhaps the most commonly used equilibrium models for the description of fertilizer chemicals, especially phosphorus, heavy metals, and pesticides. These equilibrium models include the linear and Freundlich (nonlinear) and the one- and two-site Langmuir type.

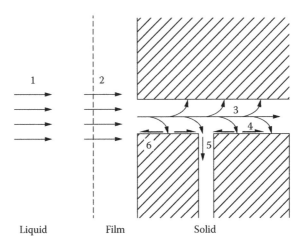

FIGURE 2.1

Transport processes in solid-liquid soil reactions. Nonactivated processes: (1) Transport in the soil solution. (2) Transport across a liquid film at the solid-liquid interface. (3) Transport in a liquid-filled micropore. Activated processes: (4) Diffusion of a sorbate at the surface of the solid. (5) Diffusion of a sorbate occluded in a micropore. (6) Diffusion in the bulk of the solid. (From C. Aharoni and Sparks, D. L. 1991. Pp. 1–18 in *Rates of Soil Chemical Processes in Soils*, edited by D. L. Sparks and D. L. Suarez. Special publication 27. Madison, WI: Soil Science Society of America. With permission.)

Soils and other geochemical systems are quite complex, and various sorption reactions are likely to occur. Such reactions are either consecutive or concurrent. Amacher (1991) and Selim and Amacher (1997) provided a schematic, shown in Figure 2.2, to illustrate several types of reactions that occur in the geochemical media of soils and the range of time for these reactions to reach equilibrium. The ion association, multivalent ion hydrolysis, and mineral crystallization reactions are all homogeneous because they occur within a single phase. The first two of these occur in the liquid phase while the last occurs in the solid phase. The other reaction types are heterogeneous because they involve transfer of chemical species across interfaces between phases. Ion association reactions refer to ion pairing, complexation (inner- and outer-sphere), and chelation-type reactions in solution. Gas-water reactions refer to the exchange of gases across the air-liquid interface. Ion exchange reactions refer to electrostatic ion replacement reactions on charged solid surfaces. Sorption reactions refer to simple physical adsorption, surface complexation (inner- and outer-sphere), and surface precipitation reactions. Mineral-solution reactions refer to precipitation/dissolution reactions involving discrete mineral phases and coprecipitation reactions by which trace constituents can become incorporated into the structure of discrete mineral phases.

Reactions in soil environments encompass a wide range of time scales as Figure 2.2 shows. Furthermore, these reactions can occur concurrently and

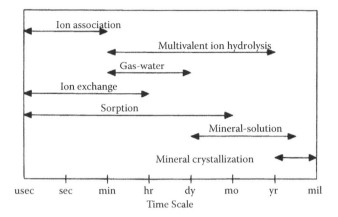

FIGURE 2.2
Time ranges required to attain equilibrium by different types of reactions in soil environments. (From H. M. Selim and Amacher, M. C. 1997. *Reactivity and Transport of Heavy Metals in Soils*. Boca Raton, FL: CRC Press. With permission.)

consecutively. The complexity of reactions in soils that occur over a time continuum defies a simple analysis of the kinetics involved (Sparks, 2003). Numerous methods have been developed to isolate and study the various types of reaction that occur in soils. The choice of method largely depends on the type of reaction to be studied, although some methods are applicable to more than one type of reaction.

Selim and Amacher (1997) developed the following sequence of steps that may be followed in a typical series of kinetic studies on heterogeneous systems such as soils:

1. Select the kinetic method(s) to be used for the reaction to be studied (batch, flow, stirred flow methods, etc.).

2. Obtain the kinetic data under varying reactant concentrations, temperature, pH, ionic strength, and composition of other solution components.

3. Determine the rate function(s) from the experimental data using initial rate, isolation, graphical, rate coefficient constancy, fractional lives, or parameter optimization methods.

4. Propose mechanism(s) from experimental rate function and other data.

5. Test mechanism(s) by conducting experiments designed to eliminate alternative mechanisms.

6. Refine or reject mechanism(s).

7. Perform additional experiments as needed to validate or eliminate revised mechanism(s).

Rarely are all these steps followed in the course of a single study and other experiments are often required to identify a mechanism(s). This is often achieved largely based on trial and error process of testing and retesting to eliminate alternate observed rate functions.

2.1 Modeling of Kinetic Sorption

Kinetic models represent slow reactions where the amount of solute sorption or transformation is a function of contact time. Most commonly encountered is the first-order kinetic reversible reaction for describing time-dependent adsorption/desorption in soils. Others include linear irreversible and non-linear reversible kinetic models. Recently, combinations of equilibrium and kinetic-type (two-site) models, and consecutive and concurrent multireaction-type models have been proposed.

2.1.1 First-Order and Freundlich Kinetics

The first-order kinetic approach is perhaps one of the earliest single forms of reactions used to describe the sorption versus time for several dissolved chemicals in soils. This may be written as:

$$\frac{\partial S}{\partial t} = k_f \left(\frac{\theta}{\rho} \right) C - k_b S \tag{2.1}$$

where the parameters k_f and k_b represent the forward and backward rates of reactions (h^{-1}) for the retention mechanism, respectively. The first-order reaction was first incorporated into the classical convection-dispersion equation by Lapidus and Amundson (1952) to describe solute retention during transport under steady-state water flow conditions. Integration of Equation 2.1 subject to initial conditions of $C = C_i$ and $S = 0$ at $t = 0$, for several C_i values, yields a system of linear sorption isotherms. That is, for any reaction time, t, a linear relation between S and C is obtained.

There are numerous examples in the literature on kinetics of pesticides and other organic sorption on various soils. Examples are shown in Figure 2.3 for imidocloprid that illustrate experimental observations where sorption over time appears linear (Jeong and Selim, 2010).

It was argued that such apparent linear behavior is not surprising for a number of reasons, including that the concentration range of the solute in solution and the adsorption optima are not attained. In addition, Selim (2011; 2012) suggested that linear behavior is also due to the uniform or homogeneous nature of the sorbing matrix. It is safe to consider organic matter as

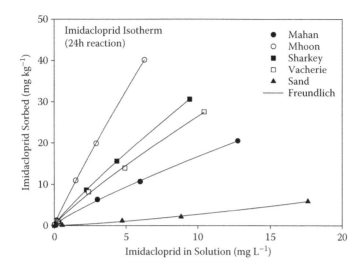

FIGURE 2.3
Adsorption isotherms for imidacloprid for five different retention soils. The solid curves are based on the Freundlich equation.

the dominant sorbent for imidoclorpid in a Vacherie soil with high organic matter. There are not many other examples of apparent linear kinetics for other solutes, with the exception of cations of low affinity such as Ca, K, and Na (Gaston and Selim, 1990a, 1990b).

Kinetic sorption that exhibits nonlinear or curve linear retention behavior is commonly observed for several reactive chemicals as depicted by the nonlinear isotherms for nickel and arsenic shown in Figures 2.4 and 2.5, respectively (Liao and Selim, 2010, and Zhang and Selim, 2005). To describe such nonlinear behavior, the single reaction given in Equation 2.1 is commonly extended to include nonlinear kinetics such that (Selim 1992):

$$\frac{\partial S}{\partial t} = k_f \left(\frac{\theta}{\rho} \right) C^b - k_b S \qquad (2.2)$$

where b is a dimensionless parameter commonly less than unity and represents the order of the nonlinear or concentration-dependent reaction and illustrates the extent of heterogeneity of the retention processes. This nonlinear reaction (Equation 2.2) is fully reversible where the magnitudes of the rate coefficients dictate the extent of kinetic behavior of retention of the solute from the soil solution. For small values of k_f and k_b, the rate of retention is slow and strong kinetic dependence is anticipated. In contrast, for large values of k_f and k_b, the retention reaction is a rapid one and should approach quasi-equilibrium in a relatively short time. In fact, at large times

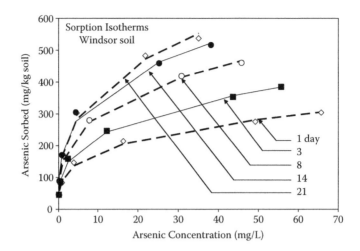

FIGURE 2.4
Adsorption isotherms for Ni on Webster soil at different retention times. The solid curves are based on the Freundlich equation.

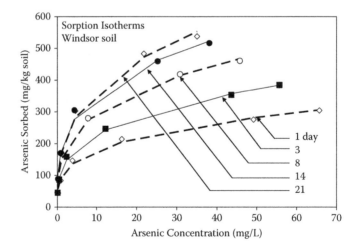

FIGURE 2.5
Adsorption isotherms for arsenic on Windsor soil at different retention times.

(i.e., as $t \to \infty$), when the rate of retention approaches zero, Equation 2.2 yields:

$$S = K_f C^b \quad where \quad K_f = \left(\frac{\theta k_f}{\rho k_b} \right) \tag{2.3}$$

Equation 2.3 is analogous to the Freundlich equilibrium equation where K_f is the solute partitioning coefficient (cm^3/g). Therefore, one may regard the parameter K_f as the ratio of the rate coefficients for sorption (forward reaction) to that for desorption or release (backward reaction).

The parameter b is a measure of the extent of the heterogeneity of sorption sites of the soil matrix. In other words, sorption sites have different affinities for heavy metal retention by matrix surfaces, where sorption by the highest-energy sites takes place preferentially at the lowest solution concentrations.

For the simple case where $b = 1$, we have the linear form:

$$S = K_d C \quad where \quad K_d = \left(\frac{\theta k_f}{\rho k_b} \right) \tag{2.4}$$

The parameter K_d is the solute distribution coefficient (cm^3/g) and of similar form to the Freundlich parameter K_f. There are numerous examples of cation and heavy metal retention, which were described successfully using the linear or the Freundlich equation (Sparks, 1989; Buchter et al., 1989). The lack of nonlinear or concentration-dependent behavior of sorption patterns as indicated by the linear case of Equation 2.1 is indicative of the lack of heterogeneity of sorption-site energies. For this special case, sorption-site energies for linear sorption processes of heavy metals may be best regarded as relatively homogeneous. A partial list of kinetic models is presented in Table 1.1 of Chapter 1.

2.1.2 Second-Order and Langmuir Kinetics

An alternative to the above first- and nth-order models is that of the second-order kinetic approach. Such an approach is commonly referred to as Langmuir kinetics and has been used for predictions of phosphorus retention and heavy metals (Selim and Amacher, 1997). Based on second-order formulation, it is assumed that the retention mechanisms are site specific where the rate of reaction is a function of the solute concentration present in the soil solution phase (C) and the number of available or unoccupied sites ϕ ($\mu g/g$ soil), by the reversible process:

$$C + \phi \underset{k_b}{\overset{k_f}{\rightleftarrows}} S \tag{2.5}$$

where k_f to k_b are the associated rate coefficients (h) and S the total amount of solute retained by the soil matrix. As a result, the rate of solute retention may be expressed as:

$$\rho \frac{\partial S}{\partial t} = k_f \, \theta \, \phi \, C - k_b \, \rho S$$

or $\hspace{9cm}$ (2.6)

$$\rho \frac{\partial S}{\partial t} = k_f \, \theta \, (S_T - S) C - k_b \, \rho S$$

where S_T (µg/g soil) represents the total amount of total sorption sites. Figure 2.6 shows the results of second-order simulations of Cu isotherm at different times during adsorption.

It is obvious that, as the sites become occupied by the retained solute, the number of vacant sites approaches zero ($\phi \rightarrow 0$) and the amount of solute retained by the soil approaches that of the total capacity of sites, that is, $S \rightarrow S_T$. Vacant specific sites are not strictly vacant. They are assumed occupied by hydrogen, hydroxyl, or other specifically sorbed species. As $t \rightarrow \infty$; that is, when the reaction achieves local equilibrium, the rate of retention becomes:

$$k_f \theta \phi C - k_b \rho S = 0, \qquad or \qquad \frac{S}{\phi C} = \left(\frac{\theta}{\rho} \right) \frac{k_f}{k_b} = \omega \qquad (2.7)$$

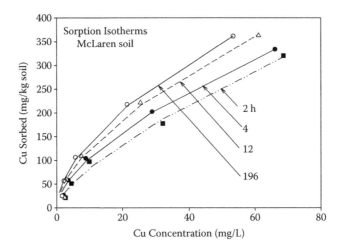

FIGURE 2.6
Adsorption isotherms for copper on McLaren soil at different retention times.

Upon further rearrangement, the second-order formulation, at equilibrium, obeys the widely recognized Langmuir isotherm equation:

$$\frac{S}{S_T} = \frac{KC}{1+KC} \tag{2.8}$$

where the parameter K ($= \theta\, k_f/k_b\,\rho$) is now equivalent to ω in Equation 2.7 and represents the Langmuir equilibrium constant. Sorption/desorption studies showed that highly specific sorption mechanisms are responsible for solute retention at low concentrations. The general view was that metal ions have a high affinity for sorption sites of oxide mineral surfaces in soils. In addition, these specific sites react slowly with reactive chemicals such as heavy metals and are weakly reversible.

2.2 Desorption and Hysteresis

Desorption of sorbed solutes from matrix surfaces is the process of detachment or release of ions or molecules to the bulk solution or the liquid phase. Knowledge of associated mechanisms is significant in understanding release or desorption behavior and provides the necessary tools for predictions and risk assessment. As with water infiltration and subsequent redistribution in the soil profile, solute release often continues for extended periods of time when compared to the duration of adsorption. Most applications of chemicals on soils and accidental spills occur over a relatively short time (hours or days). It is obvious that adsorption is dominant during such applications. This adsorption is commonly followed by extended periods of release or desorption ranging for months to years or decades.

Release or desorption is presented either as the amount sorbed or the concentration in solution versus reaction time. Results of kinetic adsorption and desorption for atrazine by sugarcane harvest residue are shown in Figure 2.7 and Figure 2.8 for a wide range of input concentrations (C_i). Such results are subsequently used to generate atrazine isotherms in the traditional manner for adsorption and desorption as shown in Figure 2.9 and Figure 2.10, respectively. It is striking that both sets (adsorption and desorption) of isotherms appear linear. Isotherm linearity for organic behavior in soils is discussed in Chapter 1.

Figure 2.11 represents an alternative to the desorption isotherms shown in Figure 2.10. This alternate method of presentation is commonly used and referred to as hysteresis, with emphasis on desorption subsequent to adsorption. The major advantage here is that two sets of separate isotherms are not required. One should note that the solid curve in the family of desorption isotherms represents that of adsorption after 504 hours, whereas all dashed

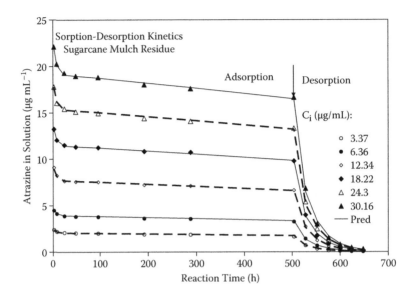

FIGURE 2.7
Atrazine concentration in solution during adsorption and desorption versus time. Results are from a batch kinetic experiment having a mulch-to-solution ratio of 1:30 and for several initial concentrations (C_i).

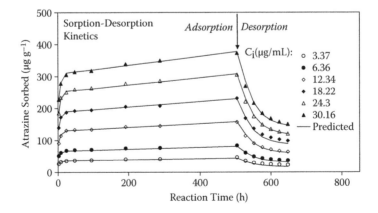

FIGURE 2.8
Amount of atrazine sorbed by sugarcane mulch residue. Results are from a batch kinetic experiment having a mulch-to-solution ratio of 1:30 and for several initial concentrations (C_i).

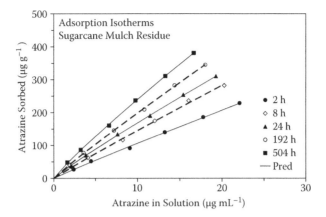

FIGURE 2.9
Time-dependent adsorption isotherms of atrazine by sugarcane mulch residue.

curves represent successive desorption isotherms. These isotherms clearly illustrate the kinetic behavior of solute sorption and that short-term isotherms (hours or days) do not necessarily provide an accurate description of their affinity to a specific type of matrix surface or a given soil. Details regarding kinetic measurements of adsorption and desorption are discussed later in this chapter (also see Selim and Zhu, 2005).

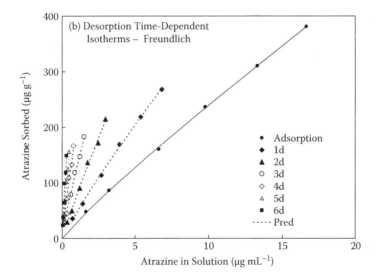

FIGURE 2.10
Time-dependent desorption isotherms of atrazine by sugarcane mulch residue. The solid line is the adsorption isotherm for 504 hours of reaction. Dashed curves are desorption isotherms.

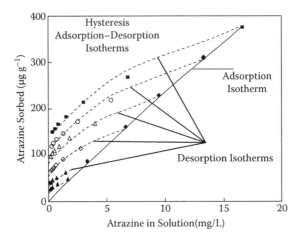

FIGURE 2.11

Traditional desorption isotherms of atrazine by sugarcane mulch residue. The solid line is the adsorption isotherm for 504 hours of reaction. Dashed curves are predictions using (a) the multireaction model (upper), and (b) the Freundlich model (lower).

The extent of hysteresis is further illustrated by the results for Zn desorption versus time for soils with distinctly different sorption affinity for Zn (see Figure 2.12). In Webster soil, desorption or release of Zn appears to be slow, indicative of strong sorption. In contrast, rapid release was observed in Windsor soil. The respective isotherms for the two soils are given in Figure 2.12 and clearly illustrate extensive hysteresis for Webster sorption consistent with strong kinetic behavior and possible irreversible reactions.

The hysteresis phenomenon has been reported for several decades in numerous studies published on colloids and colloidal chemistry. In fact, the term hysteresis is not restricted to solute sorption isotherms but is used in other disciplines, such as soil physics and hydrology. In water-unsaturated porous media, hysteresis was observed in soil-moisture content and applied suction. Discrepancies between wetting and drying curves and subsequent scanning curves give rise to the term hysteresis.

Reasons for the observed hysteresis have been discussed and various explanations have been advanced in the literature. Most center on irreversible reactions, change of phase, and formation of other ions during desorption. Selim, Davidson, and Mansell (1976) developed a mathematical proof that sorption kinetics can explain in part the discrepancies between adsorption and desorption isotherms. In fact, the explanation explicitly show that when kinetics is absent, that is, when instantaneous or equilibrium is dominant, identical adsorption and desorption isotherms are obtained. In other words, both isotherms coalesce and nonsingularity or hysteresis is not observed.

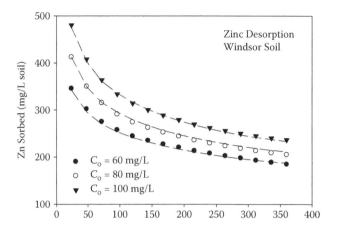

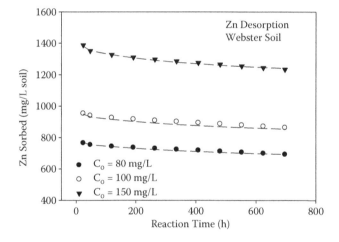

FIGURE 2.12

Zn concentration in Windsor and Webster soils versus time during desorption for various initial Zn concentrations (C_o).

To illustrate solute retention behavior (adsorption-desorption) when first-order or nonlinear kinetic (Freundlich) sorption is considered, several simulations are presented in Figures 2.14 and 2.15 (Selim, Davidson, and Mansell, 1976). As shown in Figure 2.14 (top), the linear kinetic adsorption isotherms are strongly time dependent and even after more than 50 hours only 90% of equilibrium was achieved. The slow attainment of equilibrium can be attributed primarily to the magnitude of the reaction rate coefficients k_f and k_b. Figure 2.14 (bottom) also shows simulated adsorption curves for 10

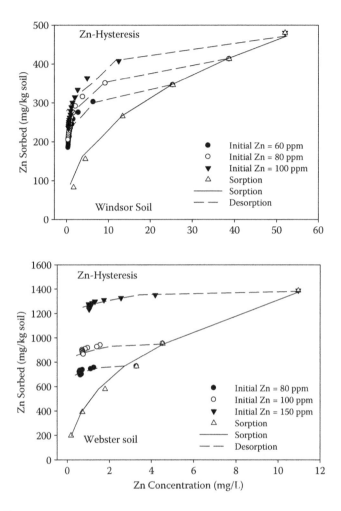

FIGURE 2.13
Adsorption and desorption isotherms for Zinc (Zn) in Windsor and Webster soils.

and 50 hours and desorption (dashed) curves initiated after 10 and 50 hours of adsorption. The simulated desorption isotherms shown were obtained by reducing the solute solution concentration one-half successively, every 10 hours (or 50 hours) until the solution concentration was less than 5 mg L^{-1}. This procedure is similar to that used in desorption studies in the laboratory. As seen from the family of (dashed) curves, desorption did not follow the same path (i.e., nonsingularity) as the respective adsorption isotherm (solid curves). Obviously, this nonsingularity or hysteresis results from failure to achieve equilibrium adsorption prior to desorption. If adsorption as

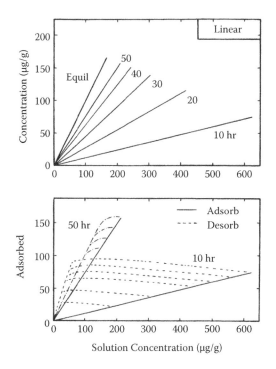

FIGURE 2.14
Simulated adsorption-desorption isotherms using a linear kinetic retention model. Desorption was initiated after 10 and 50 hours for each successive sorption step.

well as desorption were carried out for times sufficient for equilibrium to be attained, or the kinetic rate coefficients were sufficiently large, such hysteretic behavior would be minimized. For the case when nonlinear kinetic (Freundlich) sorption is considered, the respective sorption and desorption isotherms clearly illustrate hysteresis, as shown in Figure 2.15. Reaction times greater than 200 hours were required to achieve >90% equilibrium when nonlinear kinetic sorption was clear. These simulations demonstrate the concept of the influence of kinetic reaction on hysteresis regardless of the form of the sorption retention reaction.

2.2.1 Imperial Hysteresis Coefficient

Several efforts have been made to quantify hysteresis based on adsorption and desorption parameters associated with the Freundlich equation. Ma et al. (1993) defined hysteresis based on the difference between adsorption and desorption isotherms as a direct way to quantify the discrepancy between sorption and desorption. They derived the following equation as a

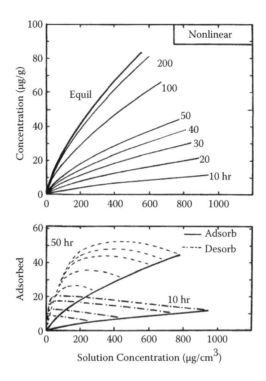

FIGURE 2.15
Simulated adsorption-desorption using a nonlinear kinetic retention model. Desorption was initiated after 10 and 50 hours for each successive sorption step.

hysteresis parameter based on the maximum difference between an adsorption and a desorption isotherm:

$$\omega = \left(\frac{N_a}{N_d} - 1 \right) \times 100 \tag{2.9}$$

where N_a and N_d are the exponent Freundlich parameter associated with adsorption and desorption, respectively. Cox, Koskinen, and Yen (1997) proposed another desorption hysteresis coefficient H, based on the ratio of desorption and adsorption isotherm parameters:

$$H = \frac{N_a}{N_d} \times 100 \tag{2.10}$$

Both coefficients ω and H are simple and easy to calculate. Zhu and Selim (2000) derived another formula to quantify the extent of hysteresis based on the area under each adsorption and desorption curve. If A_a represents the area under an adsorption isotherm curve for concentration from $C = 0$ to some solution concentration $C = C$, and A_d represents the area under a desorption isotherm at the same concentration range, we define the parameter λ as:

$$\lambda = \left(\frac{A_d - A_a}{A_a} \right) \times 100 \qquad (2.11)$$

Upon further substitution, we obtain λ as:

$$\lambda = \left(\frac{N_a + 1}{N_d + 1} - 1 \right) \times 100 \qquad (2.12)$$

Based on the above formulations, one can derive values for λ as well as ω. Selim and Zhu (2005) found that λ decreased as C_i increased for Sharkey soil, but no such relationship was observed for Commerce. Similar trends were observed for ω, whereas the opposite was observed for *H*. Ma et al. (1993) calculated ω for atrazine on Sharkey soil and indicated that ω increased linearly with incubation time, which is the time interval between the end of adsorption and the beginning of the desorption process. However, they did not observe an effect of C_i on ω. Seybold and Mersie (1996) calculated ω for metolachlor in two soils and found that ω is C_i dependent for Cullen soil that contained 31% clay and 1.3% of organic carbon, but this phenomenon was not apparent in Emporia soil, which contains less clay and less organic carbon. This dependence of desorption on C_i has been reported for other herbicides (e.g., Bowman and San, 1985; Graham and Conn, 1992, among others). It was postulated that λ increases with desorption time, indicative of dependency on the desorption history. Such behavior might be explained by the existence of irreversible reactions, which cause a decrease in desorbed herbicide amounts as desorption time increased.

2.3 Measuring Sorption

There are several experimental methods, which are described in detail in the literature, to measure sorption or affinity of a solute species to the soil matrix. Methods for measuring kinetic sorption and adsorption are available in soil chemistry and environmental soil chemistry textbooks as well as numerous scientific journals. It should be emphasized that most methods are laboratory measurements and as such suffer various shortcomings and do not depict the fate of a solute under field or *in situ* conditions of varying soil moisture or intermittent periods of wetting and drying, adsorption during solute application (or spill) and subsequent desorption, and so on. These laboratory measurements often involve mixing a few grams of soil with a volume of solution (30 mL to 2 L) having a solute of known concentration for a given equilibration time (hours to weeks).

In several methods, adsorption and desorption kinetics are quantified during flow or transport rather than where mixing is dominant. The degree of complexity of laboratory analyses and data interpretation varies extensively among the different methods. Advantages and limitation of each method and assumptions made are given in detail elsewhere (e.g., Sparks, 2003; Amacher, 1991; Selim and Amacher, 1997).

It is important to stress that data from column experiments are distinctly different from data obtained batch experiments (Miller, Sumner, and Miller, 1989; Hodges and Johnson 1987). Therefore, prediction of reactive solute transport in soil columns based on batch experiments may yield inaccurate results (Persaud, Davidson, and Rao, 1983), and it is thus recommended that kinetic retention be quantified in flow systems.

Inasmuch as batch-type experiments are commonly carried out at high solution-to-soil ratios, the data may not be applicable to transport processes (Green and Obien, 1969; Dao and Lavy, 1978). An alternative to batch experiments are short columns or thin disk flow experiments. Miller, Sumner, and Miller (1989), Hodges and Johnson (1987), and Akratanakul et al. (1983) showed that batch results underestimated the extent of ion sorption on soils and minerals when compared with short columns. The use of long columns for determination of equilibrium parameters has been practiced extensively by Schweich, Sardina, and Gaudet (1983) and Selim and Amacher (1997) for ion exchange. Furthermore, solutes that are highly reactive require a long leaching time in order to reach the input concentration at the effluent. The mixing effects in the flow direction (hydrodynamic dispersion) need to be accounted for.

In this section, selected methods of solute adsorption and desorption are described briefly. Emphasis here is on the simplicity of each method and frequency of its use.

2.3.1 One-Step Batch

This is the classical method of measuring adsorption parameters such as K_d, K_f, and Langmuir parameters. Here predried and ground soil sample (or sediment) is mixed with a tenfold volume of solution with the solute of interest in a test tube or a centrifuge tube (40 to 100 mL). The suspension is subsequently placed in a shaker for some time (hours or days). The solution is separated by filtering or centrifugation and the concentration of the solute in solution and in the solid phase is measured. This procedure is often repeated for different solute concentrations in order to provide a representative adsorption isotherm similar to those presented earlier. The range of concentrations and number of input (or initial concentrations) C_i used vary from a few points to as much as 10 to 12. Some isotherms represent a concentration range of three to four orders of magnitude (10^{-3}–10^{+2} mg/L) and some rely on one single concentration. Data analysis is simple, where the amount of solute adsorbed by the soil matrix represents the difference in concentration of the final and initial solution.

This one-step method measures adsorption at one equilibration time only. Thus, the method implicitly assumes that kinetics is not dominant and equilibrium is attained during the measurement period. Moreover, this method does not address the question of release or desorption following adsorption.

2.3.2 Kinetic Batch Reactor

This is perhaps the oldest method for measuring the rate of reactions under controlled conditions. The method is similar in principle to that described above except that it allows for repeated measurements at various times as desired to quantify solute adsorption kinetics. As illustrated in Figure 2.16,

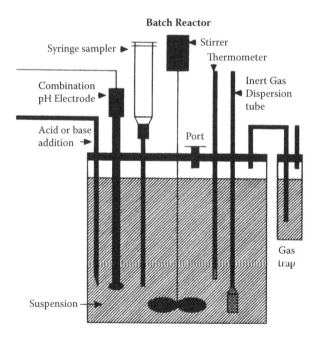

FIGURE 2.16
A typical batch reactor configuration. pH is controlled by a combination pH electrode and automatic burette connected to an auto titrator. A syringe sampler allows for removal of a subsample of suspension, an addition port permits injection of solute, an inert gas is bubbled through the suspension by means of a gas dispersion tube, and the system is vented through a gas trap. A thermometer allows for temperature monitoring, and the suspension is mixed with an overhead stirrer. (From H. M. Selim and Amacher, M. C. 1997. *Reactivity and Transport of Heavy Metals in Soils.* Boca Raton, FL: CRC Press. With permission.)

a batch reactor is normally a glass cylinder of sufficient size to obtain a sufficient number of samples to adequately describe the kinetics of the reaction (some 1 to 2 L in volume). The soil-to-solution ratio is much larger (5:100 or less) than that using the one-step reactor. As shown the suspension is mixed with an overhead stirrer or from below with a magnetic stirrer. A pH electrode is connected to maintain a constant pH for the duration of the reaction. To maintain conditions of oxidation and reduction, gas dispersion of CO_2 and O_2 in the suspension is maintained.

This batch reactor method measures adsorption at several times but is not easily adaptable to measure kinetics of release or desorption. Moreover, data analysis is a simple one, where the amount of solute adsorbed by the soil matrix represents the difference between the concentration of the sampled solution at any time t and that of the initial solution ($t = 0$). Additional details are available from Selim and Amacher (1997) and Reddy and Delaune (2008). This method is often used to assess the influence of soil parameters, notably pH and Eh (the redox potential, or rates of reaction to a specific or a combination of solute species).

2.3.3 Adsorption–Desorption Kinetic Batch Method

This method is considered one of the simplest of batch methods where solute adsorption as well desorption kinetics studies with soils and other matrices are carried out in centrifuge tubes. A schematic showing a typical adsorption-desorption experiment using centrifuge tubes is shown in Figure 2.17. The

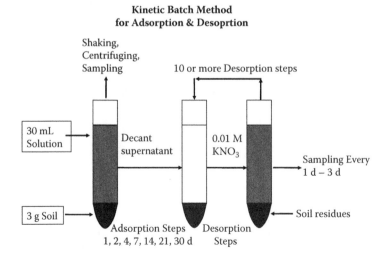

FIGURE 2.17
A schematic of batch adsorption-desorption studies in centrifuge tubes.

volume of the centrifuge tubes varies from 40 to 50 mL where a soil-to–solution ratio of 3 g to 40 mL solution (or 4 g to 40 ml solution) is used. The solution contains the solute at the desired concentration dissolved in a background solution. Commonly used are 0.005 M $CaCl_2$, KNO_3, or KCl background solution. Duplicate or triplicate samples (or tubes) are used for each initial solute concentration. For other matrices, such as organic material or plant material, 1 g of matrix to 30 or 40 mL of solution is used. The range of initial concentrations to be used varies dependent on the objectives of each study.

The mixtures or slurries in the centrifuge tubes are then shaken continuously, and after each reaction time (or sampling time), the tubes are centrifuged at 500 × g for 15 min. An aliquot is then sampled from the supernatant at the specified reaction time. The volume to be decanted can be as little as 0.2 mL if radionuclides are used. The volume of the decanted solution is normally 3 mL. The slurries are then vortex mixed and returned to the shaker after each sampling. These steps are repeated for each adsorption time. Generally, initial reaction times may vary from 2 to 4 hours followed by daily sampling for 1 to 7 days, and weekly for 4 to 6 weeks or longer. To avoid excessive changes in the soil-to-solution ratio, the number of samples should be limited to three to four if large aliquots are needed. In contrast as many as 12 or more reaction times may be carried out if radionuclides are used.

Desorption is carried out using the method of successive dilutions and commences immediately after the last adsorption time step. Each desorption step is carried out by replacing as much of the supernatant with the background solution. The amount of decanted solution and that of the background solution added must be recorded for mass balance calculations. A desorption step often consists of 24 to 36 hours of shaking for each step. Desorption is repeated for several steps as desired. The total desorption or release time depends on the number of desporption steps and the time intervals between each desorption step. Examples of adsorption-desorption results based on this method were described earlier and are presented in Figure 2.13.

The decanted solution from each adsorption and desorption step is analyzed for the solute and the amount retained by the soil matrix is calculated. Moreover, the remaining soil following the last step can be used for speciation based on different fractionation procedures (see Zhang and Selim, 2005).

2.3.4 Thin Disk Method

In this method, a thin disk of dispersed soil (1 to 2 g) deposited on a porous membrane is placed in a holder and connected to a piston-flow pump to maintain a constant flow rate (see Figure 2.18). A fraction collector is used to collect sample effluents. During adsorption, a pulse of solution containing a solute having a concentration (C_i) is introduced. The duration of the pulse varies and often is a function of sorption affinity and the

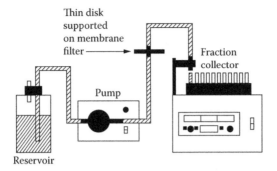

FIGURE 2.18

The thin disk flow method experimental setup. Background solute solution is pumped from the reservoir into the thin disk and the effluent is collected by a fraction collector.

flow rate. Nevertheless, the pulse is often terminated when a change in the concentration in the effluent is no longer observed. Once terminated, desorption commences using a continuous pulse of a solute-free solution. This procedure is often repeated using different initial or input concentration (C_i) in order to achieve rates of reactions over a wide range of solute concentrations.

There are several alternative ways for the use of the thin disk method to quantify solute sorption under different constraints, including ionic strength and pH. For example, pulses of increasing concentrations (C_is) may be introduced consecutively to the thin disk in order to achieve a wide range of concentrations for the same soil. Multiple solutes may also be introduced to the thin disk either concurrently (i.e., a pulse of several solutes) or consecutively where each pulse is identified by a different solute species.

There are several advantages of thin disk methods over the batch methods described above. First, the reactions associated with adsorption as well as release or desorption take place under conditions of transport or when flow is taking place. As such, thin disk methods represent an open system where incoming solute is being introduced to the soil during pulse application. The same is true during desorption or release where the incoming solution contains no solute. Another advantage is that this method closely resembles field situations compared to batch methods with their inherent artificial constraints of unrealistic ratios of soil to solution, stirring, and centrifugations.

2.3.5 Stirred-Flow Method

A schematic of the stirred-flow method is given in Figure 2.19. This method was originated by chemical engineers in the 1980s and was later adopted by

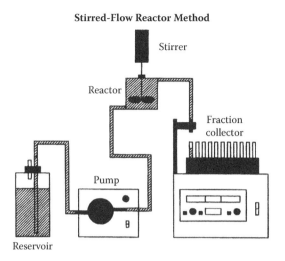

Stirred-Flow Reactor Method

FIGURE 2.19
A stirred-flow reactor connected to a reservoir of a solute through a pump. The effluent is collected by a fraction collector. (From H. M. Selim and Amacher, M. C. 1997. *Reactivity and Transport of Heavy Metals in Soils.* Boca Raton, FL: CRC Press. With permission.)

soil chemists and biogeochemists. The stirred-flow method is best regarded as a modification of the batch reactor method where reactions take place during continuous flow or transport. Selim and Amacher (1997) provided a historical perspective and modifications of the stirred-flow method as well as mass balance calculations for quantifying the amount sorbed versus time. The stirred-flow method is a combination of mixing and transport methods and is applicable for quantifying solute adsorption as well as desorption and release. Inherent advantages and disadvantage of the stirred-flow method are analogous to those of the mixing and transport methods. Sparks (2003) provided critical evaluation of this method, and of the various parameters that may be controlled (e.g., pH, Eh, etc.) during the experiment.

2.3.6 Miscible Displacement

Miscible displacement methods were originally developed by chemists and chemical engineers and are convenient to measure the transport of reactive chemicals in soils or porous media. The methods account for physical processes that govern solute transport in soils, including mass flow or convection, diffusion, and longitudinal dispersion. A primary use of these methods is not only to quantify rates of solute adsorption or desorption but also to quantify the physical characteristics governing solute transport in porous media. This is often achieved by use of a tracer or nonreactive ions such as tritium or bromide. These methods are also used by scientists dealing with transport phenomena of various reactive chemicals such as agricultural chemicals, organics (pesticides and industrial compounds), various inorganics, radionuclides, and explosives.

Since the transport of reactive solutes is directly influenced by its affinity and various chemical reactions with matrix surfaces, miscible displacement methods are best regarded as indirect means to quantify the affinity of adsorption and desorption processes in porous media. The experimental setup for miscible displacement experiments is similar to that for the thin disk method. The only exception is that a column of length L of soil or porous media is packed prior to its attachment to the pump and fraction collector as shown in Figure 2.20. Acrylic columns are commonly used and uniformly packed with air-dried soil then slowly water-saturated with a background solution at a low Darcy flux using a variable-speed piston pump. Once saturation is achieved, the flux is adjusted to the desired flow rate. A background solution is used to maintain constant ionic strength, prior to the introduction of a solute. In most studies, the application of solute pulse is maintained for an extended period. However, it is much more desirable to introduce a finite volume of the input solute pulse (i.e., for a specified duration), which is subsequently followed by solute-free solution. Introduction of free-solute solution induces release or desorption of retained solute. Solute concentrations

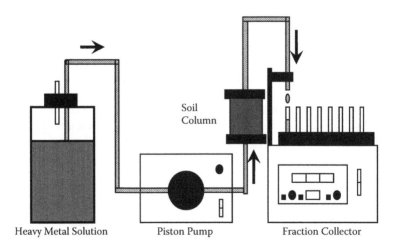

Soil
Column

Heavy Metal Solution   Piston Pump   Fraction Collector

FIGURE 2.20

Miscible displacement solute transport experimental setup. Background solution and solute are pumped upward from the reservoir through a water-saturated soil column and are collected as aliquots by the fraction collector. Separation of solid and aqueous phases is accomplished by a filter at the outlet end of the soil column.

Solute Transport in an Unsaturated Soil

FIGURE 2.21
Solute transport experimental setup for downward flow in an unsaturated soil. A tracer, solute, and background solution are pumped sequentially at flow rate as desired.

in the effluent versus time (or volume of effluent) represent the solute breakthrough curves (BTCs), which are analyzed based on various transport equations. As illustrated by the BTCs presented in the next chapter, a BTC is characterized by an adsorption or effluent (left-hand) side and desorption or release (right-hand) side. It is should be noted that the rate of reaction during release or desorption is more indicative of the fate of solute in soils than from data based on adsorption alone.

The miscible displacement methods are often to monitor solute transport during infiltration and redistribution in an initially dry soil. In other words, column transport experiments are not limited to water-saturated conditions where a steady water flow is maintained. This is illustrated in Figure 2.21, where a wetting front is advancing in a dry soil, which is monitored with depth and time of infiltration. When the wetting front reaches the lower end of the column ($z = L$), outflow commences once the pressure head reaches zero. Effluent volume is monitored with time and collected samples analyzed for various solutes. In the example shown, a short pulse of tritium (1.5 pore volumes) was first applied to a Commerce silt loam soil ($L = 30$ cm) and was subsequently followed by a large pulse of a mixture of three heavy metals (Cd, Cu, and Cr). Effluent solution was analyzed for the applied solutes with time. Moreover, after the termination of the experiment, the soil was sectioned and analyzed for the retained solutes versus depth (see Figure 2.22). Analysis of results from such column transport experiments requires solving the solute (convection-dispersion) equation and Richard's water flow equation for partially saturated soils under transient flow conditions. Such mathematical solutions are discussed in the next chapter.

FIGURE 2.22
Sectioning of a soil column after transport experiments for analysis of solute fractions retained by the soil matrix.

References

Aharoni, C., and D. L. Sparks. 1991. Kinetics of Chemical Reactions. In *Rates of Soil Chemical Processes in Soils*, edited by D. L. Sparks and D. L. Suarez, 1–18. Special publication 27. Madison, WI: Soil Science Society of America.

Akratanakul, S., L. Boersma, and O. O. Klock. 1983. Sorption Processes in Soils as Influenced by Pore Water Velocity. II. Experimental Results. *Soil Sci.* 135:331–341.

Amacher, M. C. 1991. Methods of Obtaining and Analyzing Kinetic Data. In *Rates of Soil Chemical Processes in Soils*, edited by D. L. Sparks and D. L. Suarez, 19–59. Special publication 27. Madison, WI: Soil Science Society of America.

Bowman, B.T. and W.W. Sans. 1985. Partitioning Behavior of Insecticides in Soil-Water Systems: II. Desorption Hysteresis Effects. *J. Environ. Qual.* 14:270–273.

Carrillo-Gonzalez, R., J. Simunek, S. Sauve, and D. Adriano. 2006. Mechanisms and Pathways of Trace Element Mobility in Soils. *Adv. Agron.* 91: 111–178.

Cox, L., W. C. Koskinen, and P. Y. Yen. 1997. Sorption-Desorption of Imidacloprid and Its Metabolites in Soils. *J. Agric. Food Chem.* 45: 1468–1472.

Dao, T. H., and T. L. Lavy. 1978. Atrazine Adsorption on Soil as Influenced by Temperature, Moisture Content, and Electrolyte Concentration. *Weed Sci.* 26: 303–308.

Gaston, L. A., and H. M. Selim. 1990a. Transport of Exchangeable Cations in an Aggregated Clay Soil. *Soil Sci. Soc. Am. J.* 54: 31–38.

Gaston, L. A., and H. M. Selim. 1990b. Prediction of Cation Mobility in Montmorillonitic Media Based on Exchange Selectivities of Montmorillonite. *Soil Sci. Soc. Am. J.* 54: 1525–1530.

Graham, J. S. and J. S. Conn. 1992. Sorption of Metribuzin and Metolachlor in Alaskan Subarctic Agricultural Soils. *Weed Sci.* 40:155–160.

Green R. E. and S. R. Obien. 1969. Herbicide Equilibrium in Soils in Relation to Soil Water Content. *Weed Sci.* 17:514–519.

Hodges, S. C., and G. C. Johnson. 1987. Kinetics of Sulfate Adsorption and Desorption by Cecil Soil Using Miscible Displacements. *Soil Sci. Soc. Am. J.* 50: 323–331.

Jeong, C. Y., and H. M. Selim. 2010. Modeling Adsorption-Desorption Kinetics of Imidacloprid in Soils. *Soil Sci.* 175: 214–222.

Lapidus, L., and N. L. Amundson. 1952. Mathematics for Adsorption in Beds. VI. The Effect of Longitudinal Diffusion in Ion Exchange and Chromatographic Column. *J. Phys. Chem.* 56: 984–988.

Liao, L., and H. M. Selim. 2010. Reactivity of Nickel in Soils: Evidence of Retention Kinetics. *J. Environ. Qual.* 39: 1290–1297.

Ma, L., L. M. Southwick, G. H. Willis, and H. M. Selim. 1993. Hysteretic Characteristics of Atrazine Adsorption-Desorption by a Sharkey Soil. *Weed Sci.* 41: 627–633.

Miller, D. M., M. E. Sumner, and W. P. Miller. 1989. A Comparison of Batch- and Flow-Generated Anion Adsorption Isotherms. *Soil Sci. Soc. Am. J.* 53: 373–380.

Persaud, N., J. M. Davidson, and P. S. C. Rao. 1983. Miscible Displacement of Inorganic Cations in a Discrete Homoionic Exchange Medium. *Soil Sci.*136: 269–278.

Reddy, K. R. and R. D. DeLaune. 2008. *Biogeochemistry of Wetlands—Science* and *Applications*. CRC/lewis, Boca Raton, FL

Schweich, D., M. Sardina, and J.–P. Gaudet. 1983. Measurement of Cation Exchange Isotherms from Elution Curves Obtained in a Soil Column: Preliminary Results. *Soil Sci. Soc. Am. J.* 47: 32–37.

Selim, H. M. 1992. Modeling the Transport and Retention of Inorganics in Soils. *Adv. Agron.* 47: 331–384.

Selim, H. M. 2011. *Dynamics and Bioavailability of Heavy Metals in the Rootzone*. Boca Raton, FL: CRC Press/Taylor and Francis.

Selim, H. M. 2012. *Competitive Sorption and Transport of Trace Elements in Soils and Geological Media*. Boca Raton, FL: CRC Press/Taylor and Francis.

Selim, H. M. 2013. Transport and Retention of Heavy Metal in Soils: Competitive Sorption. *Adv. Agron.* 119: 275–308.

Selim, H. M., and M. C. Amacher. 1997. *Reactivity and Transport of Heavy Metals in Soils*. Boca Raton, FL: CRC Press.

Selim, H. M., J. M. Davidson, and R. S. Mansell. 1976. Evaluation of a Two Site Adsorption-Desorption Model for Describing Solute Transport in Soils. In *Proceedings of the Summer Computer Simulation Conference*, Washington, D.C.: 444–448. Simulation Councils, La Jolla, CA.

Selim, H. M. and M. C. Amacher. 1997. *Reactivity and Transport of Heavy Metals in Soils*. CRC/lewis, Boca Raton, FL (240 p).

Selim, H. M., and H. Zhu. 2005. Atrazine Sorption–Desorption Hysteresis by Sugarcane Mulch Residue. *J. Environ. Qual.* 34: 325–335.

Seybold, C. A., and W. Mersie. 1996. Adsorption and Desorption of Atrazine, Deethylatrazine, Deisopropylatrazine, Hydroxyatrazine, and Metolachlor in Two Soils from Virginia. *J. Environ. Qual.* 25: 1179–1185.

Sparks, D. L. 1989. *Kinetics of Soil Chemical Processes*. Academic Press, San Diego, CA.

Sparks, D. L. 2003. *Environmental Soil Chemistry*, second edition. New York: Academic Press.

Sparks, D. L., and D. L. Suarez (Eds.). 1991. *Rates of Soil Chemical Processes in Soils*. Special publication 27. Madison, WI: Soil Science Society of America.

Zhang, H., and H. M. Selim. 2005. Kinetics of Arsenate Adsorption-Desorption in Soils. *Environ. Sci. Technol.* 39: 6101–6108.

Zhu, H., and H. M. Selim. 2000. Hysteretic Behavior of Metolachlor Adsorption-Desorption in Soils. *Soil Sci.* 165: 632–643.

3

Transport

To describe the general equation dealing with the transport of solutes present in the soil solution, a number of definitions must be given and the continuity or mass balance equation for the solute must be derived. One can assume that a heavy metal, or generally a solute species, may be present in a dissolved form in the soil water, that is, the solution phase. The amount of a dissolved species is expressed in terms of concentration (mass per unit volume) in the solution phase. A solute species may also be retained or sorbed by the soil matrix or be present in a precipitate or coprecipitate form.

For a given bulk volume within the soil, the total amount of solute χ (Mg cm^{-3}) for a species i may be expressed as

$$\chi_i = \Theta C_i + \rho S_i \tag{3.1}$$

where S is the amount of solute retained by the soil (Mg per gram soil), C is the solute concentration in solution (Mg cm^{-3} or mg L^{-1}), Θ is the volumetric soil water content (cm^3 cm^{-3}), and ρ is the soil bulk density (g cm^{-3}).

3.1 Continuity Equation

The continuity or mass balance equation for a solute species is a general representation of solute transport in the soil system and accounts for changes in solute concentration with time at any location in the soil. To derive the continuity equation, let us examine the transport of a solute species through a small volume element of a soil. For simplicity, we consider the volume element to be a small rectangular parallepiped with dimensions Δx, Δy, and Δz as shown in Figure 3.1. Assume that J_x is the flux or rate of movement of solute species i in the x direction, that is, the mass of solute entering the face ABCD of the volume element per unit area and time. Therefore, the solute inflow rate, or total solute mass entering into ABCD per unit time, is

$$\textit{Solute inflow rate} = J_x \, \Delta y \, \Delta z \tag{3.2}$$

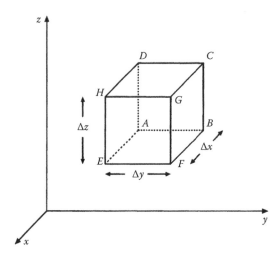

FIGURE 3.1
Rectangular volume element in the soil. (From Selim and Amacher, 1997. With permission.)

Similarly, if $J_{x+\Delta x}$ is the solute flux in the x direction for solute leaving the face EFGH, the total mass of solute leaving EFGH per unit time, that is, the solute outflow rate, is

$$Solute\ \ ouflow\ \ rate = J_{x+\Delta x}\ \Delta y \Delta z \tag{3.3}$$

From elementary calculus $J_{x+\Delta x}$ can be evaluated (approximately) from

$$J_{x+\Delta x} = J_x + \frac{\partial J_x}{\partial x} \Delta x \tag{3.4}$$

where MJ_x/Mx is the rate of change of J_x in the x direction. From Equation 3.4, the net mass of solute flow (inflow minus outflow) per unit time in the volume element from solute movement in the x direction is

$$Solute\ inflow\ rate - Solute\ outflow\ rate = (J_x - J_{x+\Delta x})\ \Delta y \Delta z$$

$$= -\frac{\partial J_x}{\partial x}\ \Delta x \Delta y \Delta z \tag{3.5}$$

Similarly, the net mass of solute flow per unit time from solute movement in the y direction is

$$-\frac{\partial J_y}{\partial y}\ \Delta x\ \Delta y\ \Delta z \tag{3.6}$$

and from solute movement in the z direction is

$$-\frac{\partial J_z}{\partial z} \Delta x \Delta y \Delta z \tag{3.7}$$

Adding Equations 3.5, 3.6, and 3.7 yields the net mass of solute (inflow-outflow) per unit time for the entire volume element as a result of solute movement in the x, y, and z directions.

$$\text{Net mass transport} = -\left(\frac{\partial J_x}{\partial x} + \frac{\partial J_y}{\partial y} + \frac{\partial J_z}{\partial z}\right) \Delta x \Delta y \Delta z \tag{3.8}$$

This net rate of solute flow represents the amount of mass of solute gained or lost within the volume element per unit time. This is often called the rate of solute accumulation. Now we assume that the solute species considered, in our example, is of the nonreactive type, that is, the solute is not adsorbed or retained by the soil matrix. Therefore, we can further assume that for a nonreactive solute S in Equation 3.1 is always zero and the solute is only present in the soil solution phase having a concentration C. Moreover, if Θ is the volumetric soil water content, that is, the volume of water per unit volume of bulk soil, then $\Theta \, \Delta x \Delta y \Delta z$ is the total volume of water in the volume element shown in Figure 3.1. At any time t, the total solute mass in the volume element is $\Theta C \, \Delta x \Delta y \Delta z$. Therefore, based on the principle of mass conservation, the rate of solute accumulation, that is, the rate of gain or loss ($\partial \Theta C / \partial t) \, \Delta x \Delta y \Delta z$, is equivalent to the net rate of mass flow (inflow-outflow). That is:

$$\frac{\partial \Theta C}{\partial t} \Delta x \Delta y \Delta z = -\left(\frac{\partial J_x}{\partial x} + \frac{\partial J_y}{\partial y} + \frac{\partial J_z}{\partial z}\right) \Delta x \Delta y \Delta z \tag{3.9}$$

By dividing both sides of Equation 3.9 by the volume element $\Delta x \, \Delta y \, \Delta z$, we have

$$\frac{\partial \Theta C}{\partial t} = -\left(\frac{\partial J_x}{\partial x} + \frac{\partial J_y}{\partial y} + \frac{\partial J_z}{\partial z}\right) = -\text{div } J \tag{3.10}$$

which is called the solute continuity equation for nonreactive solutes. For the general case where the solute is of the reactive type, we can denote the extent of solute reaction in terms of the amount retained on the soil matrix S as described in Equation (3.1). Therefore, the rate of change of the total mass χ for the ith species with time may be represented by the following general solute continuity equation for rectangular coordinates as (omitting the subscript i):

$$\frac{\partial \chi}{\partial t} = \frac{\partial(\Theta C + \rho S)}{\partial t} = -\text{div } J \tag{3.11}$$

The above equation is the general solute transport formulation dealing with the total amount of solute present in the soil system. Equation 3.11 does not include rates of production or removal of solutes from the soil, however. To achieve this, we introduce the term Q to represent a sink or a source term which accounts for the rate of solute removal (or addition) irreversibly from a unit volume of a bulk soil (Mg cm^{-3} h^{-1}). Incorporation of Q into Equation 3.11 yields:

$$\frac{\partial(\Theta C + \rho S)}{\partial t} = -\left(\frac{\partial J_x}{\partial x} + \frac{\partial J_y}{\partial y} + \frac{\partial J_z}{\partial z}\right) - Q \tag{3.12}$$

This irreversible term Q can also be considered as a rate of volatilization or a root uptake term representing the rate of extraction (Q positive) of a solute from the bulk soil or the rate of exudation of a solute (Q negative). Moreover, in the following sections, we will restrict our analysis to one-dimensional flow in the z-direction where the flux J_z is dominant.

Equation 3.12 is universally accepted as the continuity equation for solutes in the x, y, and z or in cartesian coordinates where the concentration C is presented as $C(x, y, z)$. Analogous formulations can be derived for cylindrical and spherical coordinate systems as shown in Figure 3.2. Here the concentration C is presented as $C(r, \psi, z)$ and $C(\rho, \psi, \Phi)$ for cylindrical and spherical

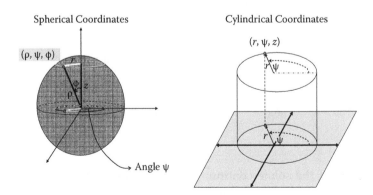

FIGURE 3.2
Spherical and cylindrical coordinate systems.

coordinates, respectively. In the figure, r, ψ, and z and ρ, ψ, and Φ represent the coordinate systems where r, ρ, and z are distances and the angles ψ and Φ are known as the azimuth and polar angles, respectively, The continuity equation in these coordinates is conveniently used in chemical and other engineering applications. In porous media, spherical coordinate systems are used in describing solute transport in ideal systems composed of spherical porous aggregates.

3.2 Transport Mechanisms

It is commonly accepted that there are two types of mechanisms that govern the transport of dissolved chemicals in soils or geological media. The first type of mechanism is an active type, which occurs regardless of whether there is water flow in the soil system. The second mechanism is a passive one and is applicable only when there is water flow. Diffusion is an active mechanism and hydrodynamic dispersion and mass transport are passive mechanisms. As discussed below, dispersion is a unique feature for porous media.

3.2.1 Mass Flow

This is often called convection and simply refers to the passive transport of a dissolved chemical with the fluid or water. That is intuitive and states that a solute moves along with the water through the soil. This mechanism is also referred to as advection or piston flow. Thus, a solute species in the soil solution and the water move together at the same flow rate:

$$U_c = q_z C \tag{3.13}$$

where U_c is the net transport of solute per unit cross-sectional area and per unit time (Mg cm^{-2} h^{-1}), q_z is the water flux density (cm^3 cm^{-2} h^{-1}) or simply the water flow velocity in the z direction. This flow velocity q_z is referred to as Darcy's flux (cm h^{-1}) and is discussed in subsequent sections.

3.2.2 Molecular Diffusion

Diffusion of ions or molecules takes place due to the random thermal motion of molecules in solution. Diffusion is an active process regardless of whether there is net water flow in the soil system. Diffusion results in a net transfer of molecules from regions of higher to lower concentrations. A common description of the diffusion is Fick's law of diffusion, that solute flux is proportional to the concentration gradient, and can be described (in the z dimension) by

$$U_d = -D_o \frac{\partial C}{\partial z} \qquad (3.14)$$

where D_o is the coefficient of molecular diffusion for a solute in a bulk or free solution (cm^2 h^{-1}). Values for D_o for several solutes are available in the literature.

To characterize diffusion of ions in soils, the diffusion coefficient given in Equation 3.14 must account for the presence of the matrix or soil particles. Specifically, the presence of soil particles results in the reduction of the volume of solution and, more importantly, the increase in the flow length of the solute. This increase in flow or path length is often referred as tortuosity. As a result, the diffusion equation for a soil system is now modified as follows (Olsen and Kemper, 1968; Nye 1979):

$$U_d = -\theta D_m \frac{\partial C}{\partial z} \qquad (3.15)$$

This new term D_m is the apparent diffusion coefficient and takes into account the effect of the solid phase of the porous medium on the diffusion. In Equation 3.15, the flux also accounts for the soil moisture content Θ. Van Schaick and Kemper (1966) and Epstein (1989) related D_m to molecular diffusion of a solute species in pure water (D_o) according to:

$$D_m = D_o \tau \qquad (3.16)$$

where τ is a tortuosity factor (dimensionless) that is defined as:

$$\tau = (\tau_o)^2 \quad \text{where} \quad \tau = L/L_a \qquad (3.17)$$

and L_a and L are the actual and shortest path lengths for diffusion and τ is the tortuosity. τ is an apparent tortuosity for porous media. This factor τ takes a value less than 1, with a range of 0.3 to 0.7 for most soils. For unsaturated soils, various forms were proposed in order to account for the soil water content Θ. Examples for expressions D_m versus D_o are presented in the classical work of Millington (1959) and Millington and Quirk (1961).

To illustrate the transport of a solute in porous medium due to diffusion, one may consider two blocks of soils, one with high and one with low concentration. At some time t, the two blocks are brought in contact with each other, and after sufficient time has elapsed, the blocks are sectioned and solute concentrations obtained. This approach yields a concentration profile versus distance from which the diffusion coefficient may be estimated. Several investigators have employed this method in the laboratory to study the effect of diffusion on solute transport in different soils (Kemper, 1986; Oscarson et al., 1992). To estimate the diffusion coefficient, the governing solute diffusion equation for nonreactive solutes must be used:

$$\frac{\partial C}{\partial t} = D\frac{\partial^2 C}{\partial z^2} \tag{3.18}$$

$$C = C_o \quad at \quad t = 0, \quad z < 0 \tag{3.19}$$

$$C = C_i \quad at \quad t = 0, \quad z < 0 \tag{3.20}$$

where C_o and C_i are the initial concentrations for blocks 1 and 2, respectively. Analytical solution of the diffusion equation is available and can be expressed as:

$$C(z,t) = C_i + 0.5(C_o - C_i) \quad erfc(z/\sqrt{4Dt}) \tag{3.21}$$

and *erfc* is the complmentary error function given by:

$$erfc(\varphi) = \frac{2}{\sqrt{\pi}} \int_{\varphi}^{\infty} e^{-r^2} d\varphi \tag{3.22}$$

Equation 3.18 is result of the incorporation of Fick's law of diffusion with the continuity equation of Equation 3.12 for the z dimension only. Equation 3.18 is commonly known as the diffusion equation in one dimension and is the subject of several books and journal articles. Solutions of Equation 3.18 for various initial conditions are available in numerous sources, including Crank (1956), Carslaw and Jaeger (1959), and Ozisik (1968).

In Figure 3.3, the concentration distributions are shown for different times after joining the two blocks. In the example it is assumed that $C_o = 1$, $C_i = 0$ mg L^{-1}, and $D = 1$ cm^2 d^{-1}. As expected as the time t for diffusion increases solute spreading from the block of high concentration increases. Similar results can be obtained for increased values of the diffusion coefficient D.

3.2.3 Dispersion

Dispersion is a highly significant transport mechanism that is unique to porous media. The term dispersion is sometimes referred to as mechanical or hydrodynamic dispersion and includes all solute-spreading mechanisms that are not attributed to molecular diffusion. The mechanical or hydrodynamic

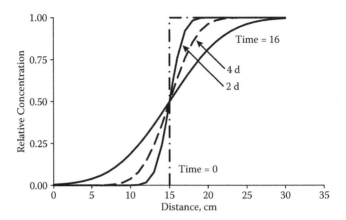

FIGURE 3.3
Effect of molecular diffusion on solute transport with distance for 0, 2, 4, and 1 d.

dispersion phenomenon is due to various factors associated with the geometry and flow characteristics of porous media. Generally, dispersion can be attributed to the nonuniform flow velocity distribution during fluid flow in porous media (Ogata, 1970, Nielsen, van Genuchten, and Biggar, 1986). In porous media the porosity results from interconnected pores having different size diameters and having different proportions. This results in different water velocities attributable to the different pore diameters or the pore size distribution of the porous medium. As a result, solutes in large pores travel to a greater distance than solutes present in small pores.

In addition, within an individual pore, according to Poiseuille's law, there is variation in velocity from the center of a pore (maximum value) to zero at the solid surface. This is illustrated in Figure 3.4, which shows solute spreading due to dispersion as well as diffusion (see Bear, 1972). According to Poiseuille's law, a velocity distribution develops over time with an increase in the advance or spreading of a solute front. Fluctuation of the flow direction or path due to tortuosity of the porous medium is another significant factor that results in solute spreading.

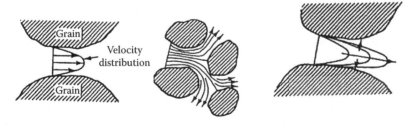

FIGURE 3.4
The spreading of a solute due to dispersion in porous media.

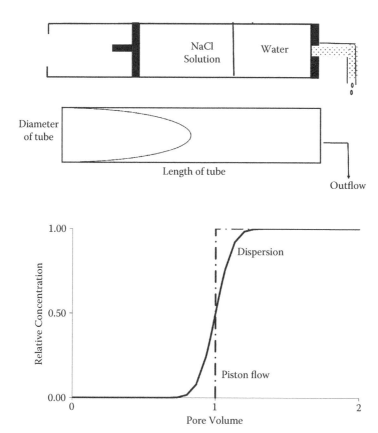

FIGURE 3.5
Piston flow, flow in a narrow tube or a soil pore, and the relative concentration versus pore volume of the effluent solution.

The effect of dispersion is that of solute spreading, which is a tendency opposite to that of the so-called piston flow. The schematics of Figure 3.5 provide a comparison of effluent concentration for piston-type flow, which is characterized by a sharp drop in concentration, that is, no solute spreading, and solute transport in a narrow tube or a soil pore. In the schematic, the increase in solute spreading is clearly manifested by the advance of the solute front over time. The change of concentration in the outflow or effluent solution versus the volume displaced clearly illustrates the difference between piston flow and that in narrow tubes or soil pores. In the schematic, the concentration is plotted versus relative volume of accumulated effluent V/V_o, where V_o (cm^3) is the total pore water volume in the soil pore (or the piston column) and V the accumulated volume of effluent (cm^3).

When dispersion is neglected, the solute moves at the same velocity and a solute front arrives as one discontinuous jump. This behavior is called *piston flow,* when a solute, if dispersion is ignored, moves in a porous medium or is displaced through the soil like a piston. Dispersion is a passive process in response to water flow. Therefore, dispersion is effective only during fluid flow, so that for a static water condition or when water flow is near zero, molecular diffusion is the dominant mixing process for solute transport in soils. A longitudinal dispersion coefficient (D_L) and transverse dispersion coefficients (D_T) are needed to describe the dispersion mechanism. Longitudinal dispersion refers to that in the direction of water flow and (D_T) refers to dispersion in directions perpendicular or transverse to the direction of flow. Longitudinal dispersive transport can be described by an equation similar to Equation 3.14 for diffusion:

$$U_m = -\Theta D_L \frac{\partial C}{\partial z} \qquad (3.23)$$

Separate studies investigated the effects of the soil water content Θ (Laryea et al., 1982; Smiles and Philip, 1978; Smiles et al., 1978) and the water flux q (Smiles and Gardiner, 1982; De Smedt and Wierenga, 1984) on D_L. A linear relationship between unique D_L and v, where v is referred to as the pore-water velocity and is given by (q/Θ), is commonly used:

$$D_L = D_o + \lambda v \qquad (3.24)$$

where D_o is the molecular diffusion coefficient in water. The term λ is a characteristic property of porous media known as the *dispersivity* (cm). Dispersivity values λ vary from a few centimeters for uniformly packed (disturbed) laboratory soil columns to several meters for field-scale experiments. Large values of λ are also reported for well-aggregated soils. In practice, an empirical parameter D rather than D_L is often introduced to simplify the flux equation (Boast, 1973). Moreover, because $D_o \ll D_L$, D_L or simply $D = \lambda v$, or a more general formula (see Bear, 1972):

$$D = D_L = \lambda v^n \qquad (3.25)$$

is often used, where n is an empirical constant with a common range of 1.0 to 1.2 (Yasuda et al., 1994; Montero et al., 1994; Jaynes, Bowman, and Rice, 1988). Therefore, D versus v may not be strictly linear and the dispersivity λ is not velocity dependent (Gerritse and Singh, 1988).

Several modifications of D vs. v are found in the literature by introducing the tortuosity coefficient τ of Equation (3.17), including that by Brusseau (1993):

$$D = D_o \tau + \lambda v^n \qquad (3.26)$$

and that of Rao et al. (1976):

$$D = (D_o + \lambda v)\tau \qquad (3.27)$$

More general regression formulas were proposed, including those of De Smedt and Wierenga (1984) and Wierenga (1977):

$$D = a + bv \qquad (3.28)$$

and

$$D = a + bv^2 \qquad (3.29)$$

where a and b are constants, with a much greater than D_o. Another regression formula is also used (Bond and Smiles, 1983; Rose, 1977; Bond, 1986):

$$D = D_m\left(\Gamma + aP_e^b\right) \qquad (3.30)$$

where I is a constant and P is the particle Peclet number defined as:

$$P = vl/D_o \qquad (3.31)$$

where the term l is a characteristic length of the porous medium.

Besides water flow velocity, D or D_o is also affected by the degree of water saturation. D is usually much larger in unsaturated soils (De Smedt and Wierenga, 1984; Laryea et al., 1982). In a horizontal infiltration study with a silty clay loam, Laryea et al. (1982) estimated increased D values as the soil water content increased. Moreover, D depends on porosity and pore size distribution. Koch and Fluhler (1993) found that D values were much larger in porous beads than in spherical solid beads.

Combining Equations 3.13, 3.14, and 3.16 yields the following simplified solute flux expression in the z direction:

$$J_z = -\Theta D \frac{\partial C}{\partial z} + q_z C \qquad (3.32)$$

which incorporates the effects of mass flow or convection as well as diffusion and mechanical dispersion. Incorporation of the flux Equation 3.19 into the conservation of mass Equation 3.13 yields the following generalized form for solute transport in soils in one dimension:

$$\frac{\partial \theta C}{\partial t} + \rho \frac{\partial S}{\partial t} = \frac{\partial}{\partial z}\left[\theta D \frac{\partial C}{\partial z}\right] - \frac{\partial q_z C}{\partial z} \qquad (3.33)$$

Equation 3.33 is commonly known as the convection-dispersion equation (CDE) for solute transport in porous media and is applicable for fully saturated and partially saturated water contents and under transient and steady flow.

For conditions where steady water flow is dominant, q and Θ are constants over space and time, that is, for uniform Θ in the soil, the simplified form of the CDE for nonreactive chemicals is

$$\frac{\partial C}{\partial t} = D\frac{\partial^2 C}{\partial z^2} - v\frac{\partial C}{\partial z} \tag{3.34}$$

where v is the pore-water velocity (q_z/Θ).

3.2.4 Equilibrium Linear and Nonlinear Sorption

Linear or linearized equilibrium sorption isotherms are the simplest form of solute retention. Here it is assumed that the amount retained by the soil (S) is related to solute concentration in solution (C) by an expression of the linear form:

$$S = K_d C \tag{3.35}$$

where K_d is the distribution coefficient ($cm^3\ g^{-1}$). This simple assumption of linear adsorption generally is valid for most pesticides and solutes of low affinity to the soils and at low concentrations. Incorporation of the linear form into the CDE yields:

$$R\frac{\partial C}{\partial t} = D\frac{\partial^2 C}{\partial z^2} - v\frac{\partial C}{\partial z} \tag{3.36}$$

where R is the retardation factor:

$$R = 1 + \frac{\rho}{\Theta} K_d \tag{3.37}$$

If there is no solute retention by the soil, K_d becomes zero and R becomes one. Such an assumption is often made for anionic and neutral tracers such as chloride, bromide, and tritium, among others. In some cases R may become less than one, indicating that only a fraction of the soil solution phase participates in the transport process. This may be the case when the solute is subject to significant anion exclusion or when relatively immobile water regions are present, for example, inside dense aggregates that do not contribute to convective transport. Van Genuchten and Weirenga (1986) suggested that, in case of anion exclusion, $(1 - R)$ may be viewed as the relative anion exclusion volume. However, for most reactive chemicals in soils, nonlinear isotherms are expected:

$$S = K_f C^b \tag{3.38}$$

which is known as the Freundlich equation. Incorporation of the linear form into the CDE yields:

$$\frac{\partial C}{\partial t} + \frac{\partial S}{\partial t} = \frac{\partial}{\partial z}\left[D\frac{\partial C}{\partial z} \right] - \upsilon\frac{\partial C}{\partial z} \tag{3.39}$$

where S is given by the Freundlich equation. The Freundlich equation does not yield an equivalent retardation factor (see Equation 3.37), which is useful in assessment of travel time and depth. Nevertheless, a nonlinear form of the retardation factor can be expressed as:

$$R = 1 + \frac{\rho}{\theta} K_f C^b \tag{3.40}$$

where some arbitrary or average concentration may be used. As discussed in Chapter 1, other sorption processes, including Langmuir, first-order, and nth-order kinetics are incorporated into the sorption term of the right-hand side of Equation 3.40.

In Figure 3.6, the movement of a solute in soils having different retardation factor R in Equation 3.37 is presented. Figure 3.6a shows a comparison of solute distribution for a nonreactive ($R = 1$) and a reactive ($R = 2$ and 4) solute, which clearly illustrates the effect of solute retention as depicted

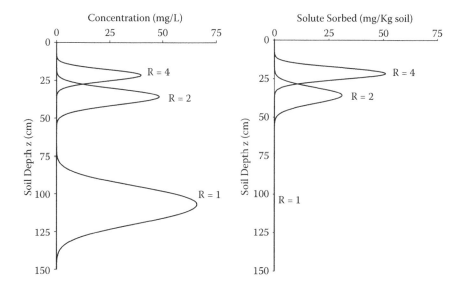

FIGURE 3.6
Solute concentration and the amount sorbed versus depth for different values of linear retardation factor R. For nonreactive solutes ($R = 1$) there was no solute sorption with depth.

by the retarded transport for the reactive solute over time. Figure 3.6b
clearly illustrates the influence of different values of R on the movement
of a solute in the soil profile. For the nonlinear case (Equation 3.40), the
influence of retardation on the movement of a solute in soils having different retardation Freundlich b values is presented in Figure 3.7. Figure 3.7a
shows a comparison of solute distribution for a linear case ($b = 1$) and a
nonlinear case ($b = 0.5$). Figure 3.7b shows a comparison of solute distribution for solutes having different values of b. For $b < 1$, the resulting
profile indicates excessive spreading, whereas for $b > 1$, the distribution is
presented by a sharp front in contrast to those where b equals 0.5 or 1. As

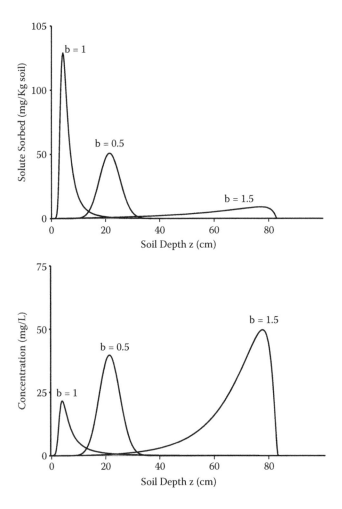

FIGURE 3.7
Solute concentration and the amount sorbed versus depth for different values of the Freundlich
parameter b. For linear sorption, b is 1.

discussed in Chapter 1, most reactive solute behavior is of the nonlinear type with $b \leq 1$.

3.3 Initial and Boundary Conditions

Solutions of the above CDE yield the concentration distribution of the amount of solute in soil solution (C) and that retained by the soil matrix (S) with time and space in soil (z, t). In order to arrive at such a solution, the initial and boundary conditions that accurately describe the experimental conditions must be specified. Several boundary conditions are identified with the problem of solute transport in porous media. First-type (Dirichlet) boundary conditions for a solute pulse input may be described as:

$$C = C_o, \quad z = 0, \quad t < t_p \tag{3.41}$$

$$C = 0, \quad z = 0, \quad t \geq t_p \tag{3.42}$$

where C_o (Mg cm^{-3}) is the concentration of the solute species in the input pulse. The input pulse application is for a duration t_p (h), which is then followed by solution that is free of solute. These boundary conditions describe a tracer solution applied at a specified rate from a perfectly mixed inlet reservoir to the surface of a finite or semi-infinite soil profile. These Dirichlet boundary conditions were used by Lapidus and Amundson (1952) and Clearly and Adrian (1973) and assume that the concentration itself can be specified at the inlet boundary. This situation is not usually possible in practice.

Third-type boundary conditions are commonly used and account for advection plus dispersion across the interface of solute at concentration C_o. For a continuous input at the soil surface, we have

$$vC_o = -D\frac{\partial C}{\partial z} + vC, \quad z = 0, \quad t > 0 \tag{3.43}$$

and for a third-type pulse-input we have

$$vC_o = -D\frac{\partial C}{\partial z} + vC, \quad z = 0, \quad t < t_p \tag{3.44}$$

$$0 = -D\frac{\partial C}{\partial z} + vC, \quad z = 0, \quad t \geq t_p \tag{3.45}$$

These conditions are for a solute or a tracer solution that is applied at a specified rate from a perfectly mixed inlet reservoir to the soil surface. Continuity of the solute flux across the inlet boundary leads directly to the above third-type boundary conditions. Third-type boundary conditions were used by Brenner (1962) and Lindstrom et al., 1967), among others.

Proper formulation of the exit boundary condition for displacement through finite laboratory columns and soils at the field scale are needed. The boundary condition at some depth L in the soil profile is often expressed as (Danckwerts, 1953):

$$\frac{\partial C}{\partial z} = 0, \quad z = L, \quad t \geq 0 \tag{3.46}$$

This second-type boundary condition is used to deal with solute effluent from soils having finite depths, such as laboratory miscible displacement columns, and describes a zero concentration gradient at $z = L$.

It is often convenient to solve the CDE where a semi-infinite rather than a finite length (L) of the soil is assumed. Under such circumstances, the appropriate condition for a semi-infinite medium is needed. Specifically, for semi-infinite systems in the field, we need a boundary condition that specifies solute behavior at large depth ($z \to \infty$). Such a boundary condition may be expressed as C = constant (commonly zero) as $z \to \infty$. An appropriate formulation of this boundary is

$$\frac{\partial C}{\partial z} = 0, \quad z \to \infty, \quad t \geq 0 \tag{3.47}$$

which is identical to that for a finite soil length (L). Kreft and Zuber (1978) and van Genuchten and Wierenga (1986) presented a discussion of the various types of boundary conditions for solute transport problems.

3.4 Exact Solutions

Analytical or exact solutions to the CDE (Equation 3.34) subject to the appropriate boundary and initial conditions are available for a limited number of situations, whereas the majority of the solute transport problems must be solved using numerical approximation methods. In general, whenever the form of the retention reaction is linear, an exact or closed-form solution is obtainable. A number of closed-form solutions are available in the literature and are compiled by van Genuchten and Alves (1982). Since several boundary conditions are commonly used, we limit our discussion to three exact solutions to the CDE (Equation 3.24) that are widely cited in the literature. All three solutions have the form:

$$C(z,t) = \begin{cases} C_i + (C_o - C_i)\, A(z,t) & 0 < t < t_p \\ C_i + (C_o - C_i)\, A(z,t) - C_o\, A(z,t-t_p) & t > t_p \end{cases} \tag{3.48}$$

where we assume a simple initial condition $(C = C_i)$ representing uniform solute concentration distribution in the soil at $t = 0$. The form of the solution in Equation 3.33 is applicable for conditions representing continuous solute application or a pulse-type input having duration t_p. For the three exact solutions, the appropriate expressions for $A(z, t)$ are given below.

3.5 Brenner Solution

Brenner (1962) considered the case of a finite soil column with the more precise third-type boundary condition at $z = 0$ that accounts for dispersion and advection across the upper surface (Equations 3.29 and 3.30). By defining the Peclet number defined in Equation 3.31

$$P = \frac{v L}{D} \tag{3.50}$$

Brenner's solution can be expressed as:

$$A(z,t) = 1 - \sum_{m=1}^{\infty} \frac{2P\beta_m \left[\beta_m \cos\left(\dfrac{\beta_m z}{L}\right) + \dfrac{P}{2}\sin\left(\dfrac{\beta_m z}{L}\right)\right] \exp\left(\dfrac{zP}{2L} - \dfrac{Pvt}{4LR} - \dfrac{\beta_m^2 vt}{PLR}\right)}{\left[\beta_m^2 + \dfrac{P^2}{4} + P\right]\left[\beta_m^2 + \dfrac{P^2}{4}\right]} \tag{3.51}$$

where the eigenvalues β_m are the positive roots of

$$\beta_m \cot(\beta_m) - \frac{\beta_m^2}{P} + \frac{P}{4} = 0 \tag{3.52}$$

Brenner's solution describes volume-averaged concentrations within the column. Because of the zero concentration gradient at $z = L$, this solution also defines a flux concentration at the lower boundary. Hence, Brenner's solution correctly interprets effluent concentrations as representing flux averaged concentrations. Another feature of Brenner's solution is that the mass balance requirement for a finite column is met. That is, the amount of solute that is entering the column minus the amount that is leaving that column equals that which is stored in the column. Brenner's solution, which is a series, converges only for relatively small values of P. In fact, Selim and Mansell (1976) found that Brenner's solution requires as many as 100 terms to obtain convergence for $P > 20$. Therefore, approximate solutions are recommended for large values of P.

3.6 Lindstrom Solution

Lindstrom et al. (1967) considered the case for a semi-infinite medium with a third-type boundary at the soil surface. Specifically, the boundary conditions given by Equations 3.29, 3.30, and 3.32 were used. This case is similar to that considered by Brenner (1962) except for a semi-infinite rather than finite column lengths. Lindstrom's solution can be expressed as:

$$A(z,t) = \frac{1}{2} \, erfc \left[\frac{Rz - vt}{(4DRt)^{1/2}} \right] + \left(\frac{v^2 t}{\pi DR} \right)^{1/2} \exp \left(-\frac{(Rz - vt)^2}{4DRt} \right)$$

$$-\frac{1}{2} \left(1 + \frac{vz}{D} + \frac{v^2 t}{DR} t \right) \exp \left(\frac{vz}{D} \right) erfc \left[\frac{Rz + vt}{(4DRt)^{1/2}} \right] \tag{3.53}$$

This solution does not suffer from convergence problems as does that of Brenner. In the meantime, it provides accurate mass balance and describes volume-averaged concentrations in the column.

3.7 Cleary and Adrian Solution

Cleary and Adrian (1973) considered a similar case to that of Brenner (1962) except for a first-type boundary condition at the inlet ($z = 0$). Their solution is for finite column lengths having boundary conditions given by Equations 3.26, 3.27, and 3.31, and may be expressed as:

$$A(z,t) = 1 - \sum_{m=1}^{\infty} \frac{2\beta_m \sin \left(\frac{\beta_m z}{L} \right) \exp \left(\frac{zP}{2L} - \frac{Pvt}{4LR} - \frac{\beta_m^2 vt}{PLR} \right)}{\left(\beta_m^2 + \frac{P^2}{4} + \frac{P}{2} \right)} \tag{3.54}$$

where the eigenvalues β_m are the positive roots of

$$\beta_m \cot(\beta_m) + \frac{P}{2} = 0 \tag{3.55}$$

This solution suffers from the same convergence problems as Brenner's, which converges only for relatively small values of P. Moreover, Cleary and Adrian's solution fails the mass-balance requirement, and also violates mass balance for the effluent curve. Van Genuchten and Weirenga (1986) recommended that this solution not be applied to solute transport column experiments. The significance of using the precise boundary conditions is illustrated by comparing the Cleary and Adrian (1973) solution with the concentration profiles calculated using the mathematical solutions of Brenner (1962) and Lindstrom et al. (1967). Unlike the boundary condition used by the other two solutions, Cleary and Adrian (1973) assumed a first-type boundary condition ($C = C_o$ at $z = 0$), which has been used by several investigators (Gupta and Greenkorn, 1973; Kirda et al., 1973; Lai and Jurinak, 1972; Warrick et al., 1971). The solution by Lindstrom et al. (1967) was developed for a semi-infinite soil column but was later applied to finite soil columns by Davidson et al. (1968) and Davidson and Chang (1972) among others.

The above solutions were used to calculate distributions of solute concentration in a 30-cm soil column (Figure 3.8) for selected times during continuous application of a solute solution (see Selim and Mansell, 1976). These concentration profiles were obtained for pore-water velocities v of 0.5, 1.5, and 3.0 cm/h. Parameters used were those of Lai and Jurinak (1972), where $D = 1.5$ cm^2 h^{-1}, $\rho = 1.30$ g cm^{-3}, $\Theta = 0.45$ cm^3 cm^{-3}, $K_d = 2.5$ cm^3 g^{-1}, and $L = 30$ cm with an R value of 8.22.

As was expected, the solution of Cleary and Adrian provided higher concentrations throughout the soil column at all times and for all three pore velocities (Figure 3.8) in comparison with results obtained by using the other two solutions (see Selim and Mansell, 1976). With decreasing pore water velocity (v) the magnitude of deviation among the three solutions increased. The higher concentrations are attributed to the assumption that $C = C_o$ at z 0 for all times. However, the concentration profiles obtained using the solution of Lindstrom et al. (1967) are essentially identical to results obtained using the Brenner (1962) solution except in the vicinity of the exit end of the soil column ($z = L = 30$ cm) at large times. Deviation of the Lindstrom et al. (1967) solution in the vicinity of $z = L$ is clearly due to forcing a semi-infinite boundary condition to describe solute transport for the finite soil columns presented here.

Solute breakthrough curves (BTCs) corresponding to pore water velocities of 0.5, 1.5, and 3.0 cm h^{-1} for the same soil parameters as in Figure 3.8 are shown in Figure 3.9 (see Selim and Mansell, 1976). BTCs are commonly used in miscible displacement studies and represent relative solute concentration C/C_o at $z = L$ versus relative volume of accumulated effluent V/V_o, where V_o is

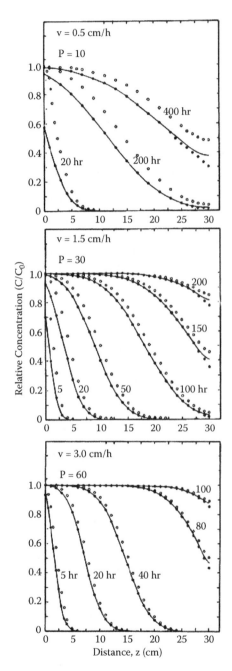

FIGURE 3.8
Relative concentration (C/C_o) versus distance z (cm) within a soil at various times for $P = 10$, 20, and 60 ($v = 0.5$, 1.5, and 3.0 cm, respectively). Solid curves were obtained using the Brenner solution, open circles using the Clearly and Adrian solution, and closed circles the Lindstrom et al. solution. (From Selim and Amacher, 1997. With permission.)

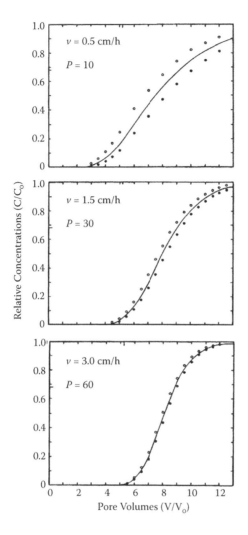

FIGURE 3.9
Relative concentration (C/C_o) versus pore volumes at $L = 30$ cm in a soil for $P = 10$, 20, and 60 ($v - 0.5$, 1.5, and 3.0 cm, respectively). Solid curves were obtained using the Brenner solution, open circles using the Clearly and Adrian solution, and closed circles the Lindstrom et al. solution. (From Selim and Amacher, 1997. With permission.)

the pore water volume in the soil column ($V_o = \Theta L$). For all three pore water velocities, the BTCs obtained using the Brenner solution occur between those obtained by using solutions of Cleary and Adrian and Lindstrom et al. As is indicated in Figure 3.7, deviation between the Brenner solution and the other two solutions decreases as the pore water velocity or Peclet number P increases. For $P = 10$, the deviation was about 8%, whereas for $P = 60$ the deviation was only 2%. Calculated BTCs for several values of the column Peclet

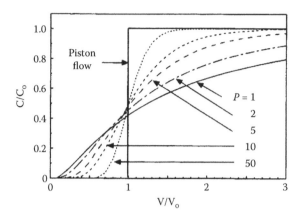

FIGURE 3.10
Relative concentration (C/C_o) versus pore volume (V/V_o) based on the Lindstrom et al. solution for $R = 1$ and $P = 1$. (From Selim and Amacher, 1997. With permission.)

number P are shown in Figure 3.10. As P decreases, the extent of spreading increases due to increased apparent dispersion coefficient D or decrease in v. Moreover, as $P \to \infty$ ($D \to 0$ or $v \to \infty$) the breakthrough resembles that for a step function at one pore volume ($V/V_o = 1$) and represents the condition of piston-type flow or convection only.

3.8 Other Exact Solutions

In addition to the exact solutions presented above for the CDE, several other exact solutions are also available. Specifically, a number of solutions are available for the CDE having the form:

$$R\frac{\partial C}{\partial t} = D\frac{\partial^2 C}{\partial z^2} - v\frac{\partial C}{\partial z} - a_1 C - a_2 \tag{3.56}$$

which includes a linear (reversible) retention term as described by the retardation factor R (of Equation 3.25). In addition, Equation 3.40 includes a first-order irreversible reaction term, with a_1 the associated rate coefficient (h^{-1}). In addition, it also includes a zero-order sink source/source term having a constant rate of loss a_2 (or gain for negative a_2). Subject to different sets of boundary conditions (Equations 3.26 to 3.32) for soil columns of finite or semi-infinite lengths, a number of exact solutions to Equation 3.40 are available in the literature (Carslaw and Jaeger, 1959; Ozisik, 1968; Selim and Mansell, 1976; van Genuchten and Alves 1982). However, most retention mechanisms are nonlinear and time dependent in nature, and analytical solutions are not available. As a result a number of numerical models using finite-difference or finite-element approximations have been utilized to solve nonlinear retention problems of multiple reactions and multicomponent solute transport for one- and two-dimensional geometries.

3.9 Estimation of D

In a number of field and laboratory miscible displacement studies, the main purpose of a tracer application is to estimate the apparent dispersion coefficient, D. A commonly used technique for estimating D is to describe tracer breakthrough results where tritium, chloride-36, bromide, or other tracers are used. It is common to use one of the above exact solutions or an approximate (numerical) solution of the CDE. In addition, a least-squares optimization scheme or curve-fitting method is often used to obtain best-fit estimates for D. One commonly used curve-fitting method is the maximum neighborhood method of Marquardt (1963), which is based on an optimum interpolation between the Taylor series method and the method of steepest descent (Daniel and Wood, 1973) and is documented in a computer algorithm by van Genuchten (1981).

The goodness-of-fit of tracer BTCs is usually unacceptable when D is the only fitting parameter. Thus, two parameters are fitted (usually D along with the retardation factor R) in order to improve the goodness-of-fit of tracer BTCs (van Genuchten, 1981). Other commonly fitted parameters include pulse duration t_p and the flow velocity v for solute retention (Jaynes, Bowman, and Rice, 1988; Andreini and Steenhuis, 1990). However, since v can be measured experimentally under steady-state flow, it may not be appropriate to fit v to achieve improved fit of the BTCs. The best-fit velocity v is often different from that measured experimentally. Estimates for R values for tritium and chloride-36 tracers are often close to unity for most soils. R greater than unity indicates sorption or simply retardation, whereas R less than one may indicate ion exclusion or negative sorption. Similar values for R for tritium and ^{36}Cl were reported by Nkedi-Kizza et al. (1983), van Genuchten and Wierenga (1986), and Selim, Schulin, and Fluhler (1987). Table 3.1 provides estimates for D obtained from tracer breakthrough results for several soils. Selected examples of measured and best-fit prediction of tritium breakthrough results for selected cases are shown in Figures 3.11 to 3.13 for two reference clays (kaolinite and monotorillonite) and a Sharkey clay soil material (Gaston and Selim, 1990b, 1991). Ma and Selim (1994) proposed the use of an effective path length L_e or a tortuosity parameter τ (L_e/L) where L_e was obtained based on mean residence time measurements. They tested the validity of fitting solute transport length (L_e) or tortuosity (τ)

TABLE 3.1

Values of the Dispersion Coefficients (*D*) for Selected Soils and Minerals from
Tritium Pulse Miscible Displacement (Soil Column) Experiments

Soils	Column Length (cm)	Pore-water Velocity (v) (cm/h)	Dispersion Coefficient (D) (cm²/h)
Sharkey Ap	15	4.00	4.96
Sharkey (6 mm)	10	1.75	9.02
Sharkey (6 mm)	15	1.85	7.66
Eustis (<2 mm)	15	.66	2.02
Cecil (<2 mm)	15	1.07	2.39
Cecil (<2 mm)	15	2.23	8.32
Cecil (<2 mm)	15	5.21	20.8
Cecil (0.5–1.0 mm)	15	2.05	3.71
Mahan (<2 mm)	15	2.02	9.79
Mahan (<2 mm)	15	3.82	15.8
Mahan (<2 mm)	15	5.29	23.0
Dothan Ap (<2 mm)	15	2.74	11.0
Dothan Bt (<2 mm)	15	2.32	11.0
Olivier (Ap)	10	2.28	1.01
Sharkey Ap	15	4.00	4.96
Mhoon	10	0.434	0.131
Sharkey (6 mm)	10	1.75	9.02
Sharkey (6 mm)	15	1.85	7.66
Vacherie	10.	0.576	1.53
Windsor (Ap)	10	1. 13	0.27
Windsor (Ap)	10	3.29	3,77
Yolo (Ap)	10	1.16	0.17
Acid wash sand	15	2.92	1.10
Glass beads + sand[a] (1:1 by weight)	15	3.08	0.73
Kaolinite + sand (1:1 by weight)	10	0.52	0.40
Bentonite + sand (1:9 by weight)	10	0.63	0.56

[a] Particle size distributions (0.25–0.50 mm, 0.50–1.0 mm, and 1–2 mm) were 0, 97.23%, and 2.13% for glass beads, and 79.96%, 19.23%, and 0 for acid washed sand, respectively.

using the CDE and concluded that L_e (or τ) can serve as a second parameter in the fitting of tracer results in addition to the dispersion coefficient (*D*). The use of an effective path length is physically more meaningful than the use of *R* as a second fitting parameter. Furthermore, independent verification of *R* values for tracer solutes when *R* is significantly different from one is not possible.

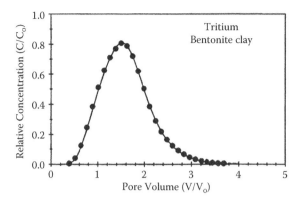

FIGURE 3.11

Tritium breakthrough results (C/C_o versus V/V_o) for a bentonite-sand column (with 1:9 clay:sand mixture). The solid curve is the fitted breakthrough curve. (From Selim and Amacher, 1997. With permission.)

Figure 3.14 represents tritium BTCs from two distinctly different soils and a reference sand material. The soils are Mahan from north Louisiana, which is predominately kaolinitic, and Webster soil from Iowa, which is of smectitic type. The reference sand was obtained from Fisher sand.

3.9.1 Numerical Solution of the Nonlinear CDE

The convection-dispersion solute transport Equation (3.24) along with the retention Equations (3.42 to 3.47) of the nonlinear multireaction approach can be solved using numerical approximations since closed-form solutions

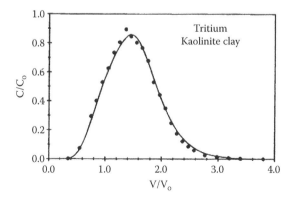

FIGURE 3.12

Tritium breakthrough results (C/C_o versus V/V_o) for a kaolinite-sand column (1:1 mixture). The solid curve is the fitted breakthrough curve. (From Selim and Amacher, 1997. With permission.)

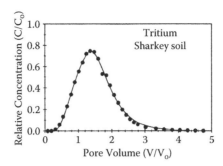

FIGURE 3.13
Tritium breakthrough results (C/C_0 versus V/V_0) for a Sharkey clay soil. The solid curve is the fitted breakthrough curve. (From Selim and Amacher, 1997. With permission.)

are not available. Commonly used numerical methods are the finite-difference explicit-implicit methods (Remson et al., 1971; Pinder and Gray, 1977). Finite-difference solutions provide distributions of solution (C) and sorbed phase concentrations (S_e, S_1, S_2, and S_3) at incremental distances Δx and time steps Δt as desired. In a finite-difference form a variable such as C is expressed as

$$C(z,t) = C(i\Delta z, j\Delta t), \qquad i = 1,2,3,\ldots,N, \quad \text{and} \quad j = 1,2,3,\ldots \qquad (3.57)$$

where

$$z = i\Delta z, \quad \text{and} \quad t = j\Delta t \qquad (3.58)$$

For simplicity the concentration $C(x, t)$ may be abbreviated as:

$$C(z,t) = C_{i,j} \qquad (3.59)$$

where the subscript i denotes incremental distance in the soil and j denotes the time step. We will assume that the concentration distribution at all incremental distances (Δx) is known for time j. We now seek to obtain a numerical approximation of the concentration distribution at time $j + 1$. The CDE (Equation 3.24) must be expressed in a finite-difference form. For the dispersion and convection terms, the finite-difference forms are

$$\Theta D \frac{\partial^2 C}{\partial z^2} = \Theta D \left(\frac{C_{i+1,j+1} - 2\,C_{i,j+1} + C_{i-1,j+1}}{2\,(\Delta z)^2} \right) +$$

$$\times \Theta D \left(\frac{C_{i+1,j} - 2\,C_{i,j} + C_{i-1,j}}{2\,(\Delta z)^2} \right) + O(\Delta z)^2 \qquad (3.60)$$

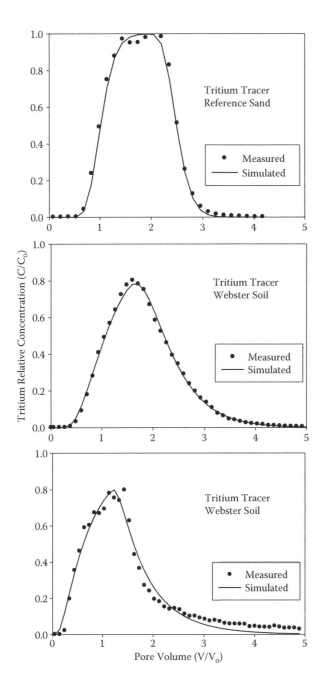

FIGURE 3.14
Tritium breakthrough results (C/C_o versus V/V_o) for a reference sand, Mahan Kaolinitic soil, and Webster smectetic soil. The solid curves are fitted breakthrough curves.

and

$$v\frac{\partial C}{\partial z} = v\left(\frac{C_{i+1,\,j+1} - C_{i,\,j+1}}{\Delta z}\right) + O(\Delta z) \tag{3.61}$$

where $O(\Delta z)^2$ and $O(\Delta z)$ are the error terms associated with the above finite-difference approximations, respectively. In Equation 3.53, the second-order derivative (the dispersion term) is expressed in an explicit-implicit form commonly known as the Crank-Nickolson or central approximation method (see Carnahan, Luther, and Wilkes, 1969). This is obtained using Taylor series expansion and is based equally on time j (known) and time $j + 1$ (unknown). Such an approximation has a truncation error, as obtained from the Taylor series expansion, in the order of $(\Delta z)^2$, which is expressed here as $O(\Delta z)^2$. In Equation 3.54, the convection term was expressed in a fully implicit form, which resulted in a truncation error of $O(\Delta z)$. In this numerical approximation, for small values of Δz and Δt, these truncation errors were assumed to be sufficiently small and were therefore ignored in our analysis (see Henrici, 1962). Chaudhari (1971) showed that due to numerical approximations, a correction to the dispersion term D is needed such that

$$D^* = D + D_n \tag{3.62}$$

where D^* is the corrected dispersion and D_n is a numerical dispersion term that is obtained from rearrangement of the higher-order terms of the Taylor series. For derivatives based on central differences, D_n is given by (Chaudhari, 1971):

$$D_n = \frac{v}{2}\left(\Delta z - \frac{v\,\Delta t}{R}\right) \tag{3.63}$$

Incorporation of D_n is a simple task and can yield significant improvement to numerical approximations.

The time-dependent term of Equation 3.24 was expressed as:

$$R\frac{\partial C}{\partial t} = R_{i,j}\left(\frac{C_{i,\,j+1} - C_{i,\,j}}{\Delta t}\right) + O(\Delta t) \tag{3.64}$$

where the retardation term R was solved explicitly as:

$$R = 1 + \left[\frac{\rho\,K_f}{\Theta}\right]b\,C^{b-1} \tag{3.65}$$

This term was also expressed in a finite difference (with iteration) as:

$$R = [R_{i,j}]^r = 1 + \frac{b\rho K_f}{\Theta} [Y^{b-1}]^r \tag{3.66}$$

where Y represents the average concentration at time j (known) and that at time $j + 1$ (unknown) for which solution is being sought, such that

$$[Y]^r = \frac{[C_{i,j+1}]^r + C_{i,j}}{2} \tag{3.67}$$

where r refers to the iteration step. Iteration is implemented here due to the nonlinearity of the equilibrium retention reaction (i.e., $b \neq 1$).

For the kinetic retention equations, the time derivative for S_1, S_2, and S_3 were also expressed in their finite-difference forms in a similar manner to the above equations. Therefore, omitting the error terms and incorporating iteration, the term associated with S_1 was expressed as:

$$\rho \frac{\partial S_1}{\partial t} \cong \Theta k_1 \left(\frac{[C_{i,j+1}]^r + C_{i,j}}{2} \right)^n - \rho k_2 [(S_1)_{i,j}]^r \tag{3.68}$$

Similarly, the kinetic term associated with S_2, when S_3 is neglected, is approximated as:

$$\rho \frac{\partial S_2}{\partial t} \cong \Theta k_3 \left(\frac{[C_{i,j+1}]^r + C_{i,j}}{2} \right)^m - \rho k_4 \left[(S_{2_{i,j}}) \right]^r \tag{3.69}$$

Moreover, the irreversible term Q was expressed in an implicit-explicit fashion as:

$$Q \cong \Theta k_s \frac{C_{i,j+1} + C_{i,j}}{2} \tag{3.70}$$

The number of iterations for the above calculations must be specified since no criteria are commonly given for an optimum number of iterations. A convenient way is based on mass balance calculations (input versus output) as a check on the accuracy of the numerical solution.

For each time step ($j + 1$), the finite difference of the solute transport equation, after rearrangement and incorporation of the initial and boundary conditions in their finite-difference form, can be represented by a set of N equations having N unknown concentrations. The form of the N equations is

$$a_{i,j} C_{i-1,j+1} + b_{i,j} C_{i,j+1} + u_{i,j} C_{i+1,j+1} = e_{i,j} \qquad (3.71)$$

where N is the number of incremental distances in the soil ($N = L/\Delta x$). By including the appropriate initial and boundary conditions in their finite-difference forms, Equation 3.33 can be written in matrix-vector notation as:

$$B(\overrightarrow{h})^{n+1} = \overrightarrow{w} \qquad (3.72)$$

where B is a tridiagonal real matrix and h and w denote the associated real column vectors (the arrows indicate vectors). The matrix B may be written as:

$$
\begin{bmatrix}
b^n_1 & u^n_1 & & & & & \\
a^n_2 & b^n_2 & u^n_2 & & & & \\
 & a^n_3 & b^n_3 & u^n_3 & & & \\
 & & & - & & & \\
 & & & & - & & \\
 & & & & & - & \\
 & & & & a^n_{N-1} & b^n_{N-1} & u^n_{N-1} \\
 & & & & & a^n_N & b^n_N
\end{bmatrix}
\qquad (3.73)
$$

The coefficients of the main diagonal of the matrix B, in absolute values, are greater than the raw sum of the off-diagonal coefficients. Hence, the matrix B is strictly diagonally dominant. Therefore, the matrix is nonsingular, and there exists a solution $n + 1$ for the matrix vector equation (Equation 3.71) that is unique. The tridiagonal system of equations was solved by an adaptation of the Gaussian algorithm. Therefore, for any time step, $n + 1$, where all variables are assumed to be known at time step n, Equation 3.71 can be solved sequentially until a desired time t is reached.

The coefficients a, b, u, and e are the associated set of equation parameters. The above N equations were solved simultaneously, for each time step, using the Gaussian elimination method (Carnahan, Luther, and Wilkes, 1969) in order to obtain the concentration C at all nodal points (i) along the soil profile. Solution for a set of linear equations such as the Thomas algorithm for tridiagonal matrix-vector equations can be used (Pinder and Gray, 1977). The newly calculated C values can be used subsequently as input values in the solution for solute retention. The solution of these equations thus provides the amount of sorbed phases due to the irreversible and reversible reactions at the same time $(j + 1)$ and all incremental distances along the soil profile.

3.9.2 Transport versus Batch

In most experimental investigations dealing with the fate of chemicals in soils or porous media transport is absent. For example, methods for quantifying adsorption isotherm for a range of concentrations and over time are rarely conducted under flow conditions. A list of such methods where net transport of a solute is absent is discussed in Chapter 1.

When transport is absent, the problem is simplified considerably and becomes that of an initial-value-problem where space is ignored. Simple analytical solutions are available for the linear case (Equation 3.56). However, for Freundlich (Equation), Langmuir (Equation), or other nonlinear formulations to describe the retention reaction processes, iterations must be used to solve the initial-value-problem. The solution becomes that representing no-flow batch conditions where the retention is to be described over time. The major exception between the above formulations and that excluding transport is in the way the equilibrium sorbed phase concentration (S_e) is calculated. For any given time step j, the amount in soil solution C and that in the sorbed phase S_e are in local equilibrium and their amounts are related by the K_f value according to the nonlinear Freundlich equation (3.42). Therefore, the total amount in the solution and sorbed phases (S_e) is

$$H = \Theta C + \rho S_e = \Theta C + \rho K_f C^b, \qquad t \geq 0 \tag{3.74}$$

As a result, one can calculate, from C and S_e, the amount H at any time step j. Now to estimate these variables at time step $j + 1$, subsequent to the calculations of all other variables (i.e., S_1, S_2, etc.), one can calculate a new value for H and partition such a value between C and S_e (based on the Freundlich equation) using the following expression:

$$C = \frac{H}{\Theta + \rho K_f C^{b-1}} \tag{3.75}$$

which is derived directly from Equation 3.74, which is based on the newly calculated H for the sum of concentration and equilibrium sorbed phases. Equation

3.75 is an implicit equation for C and where iteration is necessary. Specifically, a solution for C at each time step j and iteration step r can be obtained as follows:

$$[C_{i,j+1}]^{r+1} = \frac{H}{\Theta + \rho K_f \left([C_{i,j+1}]^r\right)^{b-1}} \qquad (3.76)$$

All other finite-difference expressions are similar to those described above.

3.9.3 Simulations

In order to illustrate the kinetic behavior of reactive solutes having various retention mechanisms, the transport model was used to provide a number of simulations. Figures 3.14 through 3.20 are selected simulations that illustrate the sensitivity of solution concentration results to a wide range of parameters associated with equilibrium as well as kinetic retention reactions. The soil parameters selected for these illustrations are: $\rho = 1.25$ g cm^{-3}, $\Theta = 0.4$ cm^3, $L = 10$ cm, $C_i = 0$, $C_o = 10$ mg L^{-1}, and $D = 1.0$ cm^2 h^{-1}. Here it was assumed that a solute pulse was applied to a fully water-saturated soil column initially devoid of solute. In addition, a steady water-flow velocity (q) is assumed constant, with a Peclet number P (= $qL/\Theta D$) of 25. The length of the pulse was assumed to be three pore volumes, which was then followed by several pore volumes of a solute-free solution.

The influence of the Freundlich distribution coefficient K_f of the nonlinear equilibrium type reaction on the transport of dissolved chemicals is shown in Figure 3.15. The shape of the BTCs reflects the influence of nonlinear

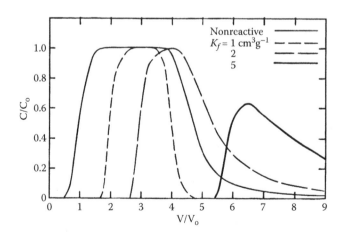

FIGURE 3.15
Breakthrough curves for several Freundlich K_f values and $b = 0.5$. (From Selim and Amacher, 1997. With permission.)

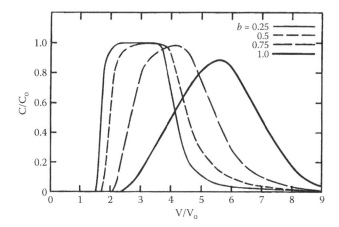

FIGURE 3.16
Breakthrough curves for several values of $b < 1$. (From Selim and Amacher, 1997. With permission.)

equilibrium Freundlich-type sorption. For the nonreactive case ($K_f = 0$), which indicates no solute retardation, the sorption (or effluent) side and the desorption side of the BTC appear symmetric. Here the solute concentration (C/C_o) slightly exceeds 0.5 for V/V_o of 1. As the Freundlich distribution coefficient K_f increased, the solute became more retarded as is clearly illustrated by the location of the sorption side of the BTCs. For example, for the case

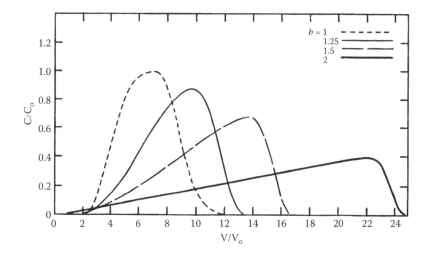

FIGURE 3.17
Breakthrough curves for several Freundlich values of $b > 1$. (From Selim and Amacher, 1997. With permission.)

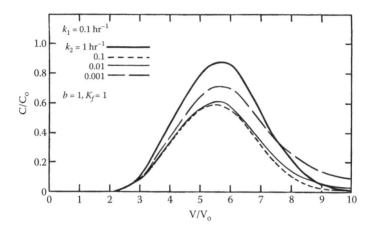

FIGURE 3.18
Breakthrough curves for several values of the kinetic rate coefficients k_1 and k_2. (From Selim and Amacher, 1997. With permission.)

where $K_f = 2$, nearly three pore volumes were required prior to the detection of solute in the effluent solution. In the meantime, a reduction of concentration maxima and the presence of tailing of the desorption side is observed for large K_f values. This is due not only to the large K_f values used but also to the nonlinearity of the equilibrium mechanism ($b \neq 1$) chosen here.

The influence of a wide range of b values on the shape of the BTC is shown in Figures 3.16 and 3.17. For all the BTCs shown in Figures 3.16 and 3.17, a K_f of unity was used. For values of $b < 1$, the shape of the BTCs indicates a sharp

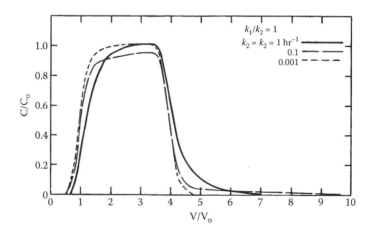

FIGURE 3.19
Breakthrough curves for several b values where the kinetic rate coefficients k_1 and k_2 where the ratio k_1/k_2 remains invariant. (From Selim and Amacher, 1997. With permission.)

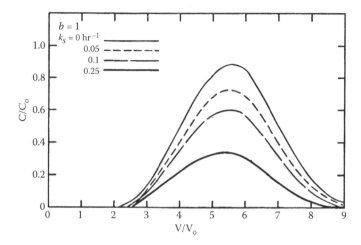

FIGURE 3.20
Breakthrough curves for several values of the rate coefficient k_s, where $b = 1$. (From Selim and Amacher, 1997. With permission.)

rise in concentration or a steep sorption side with an increase of the tailing of the desorption side for decreasing values of b. In contrast, for $b > 1$ the sorption side indicates a slow increase of the associated concentration and there is a lack of tailing of the desorption side of the BTCs.

The significance of the rate coefficients k_1 and k_2 associated with a kinetic reaction on solute retention and transport may be illustrated by the BTCs shown in Figures 3.18 and 3.19 where a range of rate coefficients differing by three orders of magnitude were chosen. For the BTCs shown in Figure 3.18, the forward rate coefficients were constant ($k_1 = 0.10$ h^{-1}), whereas k_2 varied from 1 to 0.001 h^{-1}. A decrease in concentration maxima and a shift of the BTCs resulted as the value for k_2 decreased. Such a shift of the BTCs signifies an increase in solute retention due to the influence of the kinetic mechanism associated with S_1. As the rate of backward reaction (k_2) decreases or k_1/k_2 increases, the amount of S_1 retained increases and solute mobility in the soil becomes more retarded. The BTCs shown in Figure 3.19 illustrate the significance of the magnitude of the kinetic rate reactions k_1 and k_2 while the ratio k_1/k_2 remained constant. It is obvious that, as the magnitude of the rate coefficients increased, the amount of solute retained increased and an increased solute retardation becomes evident. Moreover, for extremely small k_1 and k_2 values (e.g., of 0.001 h^{-1}), the BTC resembles that for a nonreactive solute due to limited contact time for solute retention by the soil matrix under the prevailing water-flow velocity conditions. On the other hand, large rate coefficients are indications of fast or instantaneous retention reactions. Specifically, rapid reactions indicate that the retention process is less kinetic and approaches equilibrium conditions in a relatively short contact time.

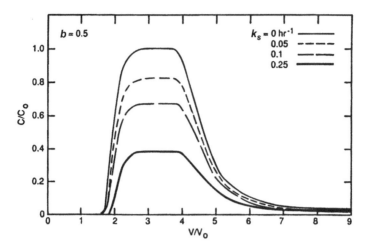

FIGURE 3.21
Breakthrough curves for several values of the rate coefficient k_s, where $b = 1$. (From Selim and Amacher, 1997. With permission.)

In the BTCs shown in Figure 3.19, the irreversible retention mechanism for solute removal via the sink term was ignored (i.e., $Q = 0$). The influence of the irreversible kinetic reaction is straightforward, as shown in Figures 3.20 and 3.21. This is manifested by the lowering of the solute concentration for the overall BTC for increasing values of k_s. Since a first-order irreversible reaction was assumed for the sink term, the amount of irreversibly retained solute and thus lowering of the BTC is proportional to the solution concentration. The primary difference between the BTCs shown in Figures 3.20 and 3.21 is due to the value of the nonlinear parameter b associated with the Freundlich equilibrium retention mechanism. In Figure 3.20, $b = 1$, whereas $b = 0.5$ for the simulations shown in Figure 3.21. All other parameters remained invariant ($k_1 = 0.001$, $k_2 = 0.01$ h^{-1}, and $K_f = 1$ cm^3/g).

References

Andreini, M. S., and T. S. Steenhuis. 1990. Preferential Paths of Flow under Conventional and Conservation Tillage. *Geoderma*. 46: 85–102.

Bear, J. 1972. *Dynamics of Fluids in Porous Media*. New York: Elsevier.

Bond, W. J. 1986. Velocity-Dependent Hydrodynamic Dispersion During Unsteady, Unsaturated Soil Water Flow: Experiments. *Water Resour. Res.* 22: 1881–1889.

Bond, W. J. and D. E. Smiles. 1983. Influence of Velocity on Hydrodynamic Dispersion during Unsteady Soil Water Flow. *Soil Sci. Soc. Am. J.* 47: 438–441.

Brenner, H. 1962. The Diffusion Model of Longitudinal Mixing in Beds of Finite Length: Numerical Values. *Chem. Eng. Sci.* 17:220–243.

Brusseau, M. L. 1993. The Influence of Solute Size, Pore Water Velocity, and Intraparticle Porosity on Solute Dispersion and Transport in Soil. *Water Resour. Res.* 29: 1071–1080.

Carnahan, P., H. Luther, and J. O. Wilkes. 1969. *Applied Numerical Methods.* New York: Wiley.

Carslaw, H. S., and J. C. Jaeger. 1959. *Conduction of Heat in Solids.* Oxford, United Kingdom: Clarendon Press.

Chaudhari, N. M. 1971. An Improved Numerical Technique for Solving Multi-Dimensional Miscible Displacement Equations. *Soc. Petr. Eng. J.* 11: 277–284.

Cleary, R. W., and D. D. Adrian. 1973. Analytical Solution of the Convective-Dispersive Equation for Cation Adsorption. *Soil Sci. Soc. Am. Proc.* 37: 197–199.

Crank, J. 1956. *The Mathematics of Diffusion.* Oxford, United Kingdom: Clarendon Press.

Danckwerts, O. V. 1953. Continuous Flow Systems. *Chem. Eng. Sci.* 2: 1–13.

Daniel, C. and F. S. Wood. 1973. *Fitting Equations To Data.* Wiley-Interscience, New York.

Davidson, J. M. and R. K. Chang. 1972. Transport of Picloram in Relation to Soil Physical Conditions and Pore-Water Velocity, *Soil. Sci. Soc. Am. Proc.* 36: 257–261.

Davidson, J. M., C. E. Rieck, and P. W Santelmann. 1968. Influence of Water Flux and Porous Material on the Movement of Selected Herbicides. *Soil Sci. Soc. Am. Proc.* 32: 629–633.

De Smedt, F., and P. J. Wierenga. 1984. Solute Transfer through Columns of Glass Beads. *Water Resour. Res.* 20: 225–232.

Epstein, N. 1989. On Tortuosity and the Tortuosity Factor in Flow and Diffusion through Porous Media. *Chem. Eng. Sci.* 44: 777–780.

Gaston, L. A., and H. M. Selim. 1990. Prediction of Cation Mobility in Montmorillonitic Media Based on Exchange Selectivities of Montmorillonite. *Soil Sci. Soc. Am. J.* 54: 1525–1530.

Gaston, L. A., and H. M. Selim. 1991. Predicting Cation Mobility in Kaolinite Media Based on Exchange Selectivities of Kaolinite. *Soil Sci. Soc. Am. J.* 55: 1255–1261.

Gerritse, R. G. and R. Singh. 1988. The Relationship Between Pore Water Velocity and Longitudinal Dispersivity of Cl, Br, and D_2O in Soils. *J. Hydrol.* 104: 173–180.

Gupta, S. P. and R. A. Greenkorn. 1973. Dispersion during Flow in Porous Media with Bilinear Adsorption, *Water Resour. Res.* 5: 1357–1368.

Henrici, P. 1962. *Discrete Variable Methods in Ordinary Differential Equations.* New York: Wiley.

Jaynes, D. B., R. S. Bowman, and R. C. Rice. 1988. Transport of a Conservative Tracer in a Field Under Continuous Flood Irrigation. *Soil Sci. Soc. Am. J.* 52: 618–624.

Kemper, W. D. 1986. Solute Diffusivity. In *Methods of Soil Analysis*, Volume I: *Physical and Mineralogical Methods*, edited by A. Klute, 1007–1024. Madison, WI: Soil Science Society of America.

Kirda, C., D. R. Nielsen, and J. W. Biggar. 1973. Simultaneous transport of chloride and Water during Infiltration. *Soil Sci. Soc. Ame. Proc.* 37: 339–345.

Kreft, A., and A. Zuber. 1978. On the Physical Meaning of the Dispersion Equation and Its Solution for Different Initial and Boundary Conditions. *Chem. Eng. Sci.* 33: 1471–1480.

Koch, S. and H. Flühler. 1993. Non-Reactive Solute Transport with Micropore Diffusion in Aggregated Porous Media Determined by a Flow-Interruption Method. *J. Contam. Hydrol.* 14: 39–54.

Lai, S.-H., and J. J. Jurinak. 1972. Cation Adsorption in One Dimensional Flow through Soils: A Numerical Solution. *Water Resour. Res.* 8: 99–107.

Laryea, K. B., D. E. Elrick, and M. J. L. Robin. 1982. Hydrodynamic Dispersion Involving Cationic Adsorption during Unsaturated, Transient Water Flow in Soil. *Soil Sci. Soc. Am. J.* 46: 667–671.

Lindstrom, F. T., R. Haque, V. H. Freed, and L. Boersma. 1967. Theory on Movement of Some Herbicides in Soils: Linear Diffusion and Convection of Chemicals in Soils. *Environ. Sci. Technol.* 1: 561–565.

Lapidus, L., and N. L. Amundson. 1952. Mathematics for Adsorption in Beds. VI. The Effect of Longitudinal Diffusion in Ion Exchange and Chromatographic Column. *J. Phys. Chem.* 56: 984–988.

Ma, L. and H. M. Selim. 1994. Tortuosity, Mean Residence Time and Deformation of Tritium Breakthroughs from Uniform Soil Columns. *Soil Sci. Soc. Am. J.* 58: 1076–1085.

Millington, R. J. 1959. Gas Diffusion of Porous Solids. *Science* 30: 100–102.

Millington, R. J., and J. P. Quirk. 1961. Permeability of Porous Solids. *Trans. Farad. Soc.* 57: 1200–1206.

Montero, J. P., J. O. Munoz, R. Abeliuk, and M. Vauclin. 1994. A Solute Transport Model for the Acid Leaching of Copper in Soil Columns. *Soil Sci. Soc. Am. J.* 58: 678–686.

Marquardt, D. W. 1963. An Algorithm for Least-Squares Estimation of Non-Linear Parameters. *J. Soc. Ind. Appl. Math.* 11: 431–441.

Nielsen, D. R., M. Th. van Genuchten, and J. M. Biggar. 1986. Water Flow and Transport Processes in the Unsaturated Zone. *Water Resour. Res.* 22: 89S–108S.

Nkedi-Kizza, P., J. W. Biggar, M. Th. van Genuchten, P. J. Wierenga, H. M. Selim, D. M. Davidson, and D. R. Nielsen. 1983. Modeling Tritium and Chloride 36 Transport through an Aggregated Oxisol. *Water Resour. Res.* 19: 691–700.

Nye, P. H. 1979. Diffusion of Ions and Uncharged Solutes in Soils and Soil Clays. *Adv. Agron.* 31: 225–272.

Ogata, A. 1970. *Theory of Dispersion in Granular Medium.* U.S. Geological Survey. Professional Paper No. 411–1.

Olsen, S. R., and W. D. Kemper. 1968. Movement of Nutrients to Plant Roots. *Adv. Agron.* 20:91–151.

Oscarson, D. W., H. B. Hume, N. G. Sawatsky, and S. C. H. Cheung. 1992. Diffusion of Iodide in Compacted Bentonite. *Soil Sci. Soc. Am. J.* 56: 1400–1406.

Ozisik, M. N. 1968. *Boundary-value problems of heat conduction.* Scranton, PA: International Textbook Company.

Pinder, G. F. and W Gray, 1977. *Finite Element Simulation in Surface and Subsurface Hydrology,* Academic Press, New York.

Rao, P. S. C., R. E. Green, L. R. Ahuja, and J. M. Davidson. 1976. Evaluation of a Capillary Bundle Model for Describing Solute Dispersion in Aggregated Soils. *Soil Sci. Soc. Am. J.* 40: 815–820.

Remson, I., G. M. Hornberger, and F. J. Molz. 1971. *Numerical Methods in Subsurface Hydrology,* Wiley, New York.

Rose, D. A. 1977. Hydrodynamic Dispersion in Porous Materials. *Soil Sci.* 123: 277–283.

Selim, H. M. and M. C. Amacher. 1997. *Reactivity and Transport of Heavy Metals in Soils.* CRC, Boca Raton, FL (240 p).

Selim, H. M., and R. S. Mansell. 1976. Analytical Solution of the Equation of Reactive Solutes through Soils. *Water Resour. Res.* 12: 528–532.

Selim, H. M., R. Schulin, and H. Fluhler. 1987. Transport and Ion Exchange of Calcium and Magnesium in Aggregated Soil. *Soil Sci. Soc. Am. J.* 5l: 876–884.

Smiles, D. E., and J. R. Philip. 1978. Solute Transport During Absorption of Water by Soil: Laboratory Studies and Their Practical Implications. *Soil Sci. Soc. Am. J.* 42: 537–544.

Smiles, D. E., J. R. Philip, J. H. Knight, and D. E. Elrick. 1978. Hydrodynamic Dispersion during Absorption of Water by Soil. *Soil Sci. Soc. Am. J.* 42: 229–234.

Smiles, D. E., and B. N. Gardiner. 1982. Hydrodynamic Dispersion during Unsteady, Unsaturated Water Flow in a Clay Soil. *Soil Sci. Soc. Am. J.* 46: 9–14.

van Genuchten, M. Th. 1981. Non-Equilibrium Transport Parameters from Miscible Displacement Experiments. Research Report No. 119, U.S. Salinity Lab., Riverside, CA.

van Genuchten, M. Th., and P. J. Wierenga. 1977. Mass Transfer Studies in Sorbing Porous Media. II. Experimental Evaluation with Tritium (^3H_2O). *Soil Sci. Soc. Am. J.* 41: 272–277.

van Genuchten, M. Th., and P. J. Wierenga. 1986. Solute Dispersion Coefficients and Dispersion Factors. In *Methods of Soil Analysis*, Part 1, edited by A. Klute, 1025–1054. Agronomy Monograph no. 9 (2nd ed), Madison, WI: American Soil Association.

van Schaik, J. C., and W. D. Kemper. 1966. Chloride Diffusion in Clay-Water System. *Soil Sci. Soc. Am. Proc.* 30: 22–25.

Warrick, A. W., J. W. Biggar, and D. R. Nielsen. 1971. Simultaneous Solute and Water Transfer for an Unsaturated Soil. *Water Resour. Res.* 7: 1216–1225.

Wierenga, P. J. 1977. Solute Distribution Profiles Computed with Steady-State and Transient Water Movement Models. *Soil Sci. Soc. Am. J.* 41: 1050–1055.

Yasuda, H., R. Berndtsson, A. Bahri, and K. Jinno. 1994. Plot-Scale Solute Transport in a Semiarid Agricultural Soil. *Soil Sci. Soc. Am. J.* 58: 1052–1060.

4

Effect of Scale

The convection-dispersion equation (CDE) is often used to describe solute transport in geologic systems under saturated and unsaturated conditions. Dispersivity, one of the parameters of the CDE, is a measure of the dispersive properties of a geologic system. Traditionally, it has been considered as a characteristic single-valued parameter for an entire medium (Pickens and Grisak, 1981b). However, a number of studies have shown that a constant dispersivity is not always adaquate, and a dispersivity that is dependent on the mean travel distance and/or scale of the geologic system is often needed (Fried, 1972; Sudicky and Cherry, 1979; Pickens and Grisak, 1981a; Gelhar, Welty, and Rehfeldt, 1992; Khan and Jury, 1990).

The dependence of dispersivity on the mean travel distance and/or scale of the geologic system is referred to as the "scale effects." Pickens and Grisak (1981a) provided a detailed review of the scale effects in field dispersion investigations. They summarized results from several computer simulations and laboratory and field transport studies. They found that dispersivities obtained from computer modeling studies of contamination zones ranged from 12 to 61 m and tend to increase with the scale of the contamination zone (Table 1 in Pickens and Grisak, 1981a). In contrast, dispersivities obtained from the analysis of laboratory breakthrough curve (BTC) data on repacked materials were of the order of 0.01 to 1.0 cm, and those obtained from analysis of various types of field tracer tests ranged between 0.012 and 15.2 m (Table 2 in Pickens and Grisak, 1981a). Gelhar, Welty, and Rehfeldt (1992) provided a critical review of some 104 dispersivity values determined from 59 different sites. The longitudinal dispersivities ranged from 10^{-2} to 10^4 m for scales ranging from 10^{-1} to 10^5 m. Although fairly scattered, the data indicated a trend of increase in the longitudinal dispersivity with observation scale.

Case studies conducted by Peaudecerf and Sauty (1978) showed that the dispersivity changes with distance. From a field transport experiment, Sudicky and Cherry (1979) found that dispersivity values for chloride based on analytical solutions increased with mean travel distance in the groundwater flow domain. Fried (1972) reported longitudinal dispersivities from several sites. He reported several values ranging from 0.1 to 0.6 m for the local (aquifer stratum) scale, 5 to 11 m for the global (aquifer thickness) scale, and 12.2 m for the regional (several kilometers) scale (cited in Pickens and

Grisak, 1981a). Later, Fried (1975) defined several scales in terms of the "mean travel distance" of a tracer or contaminant. These scales are:

- *Local scale*, ranging between 2 and 4 m
- *Global scale 1*, ranging between 4 and 20 m
- *Global scale 2*, ranging between 20 and 100 m
- *Regional scale*, larger than 100 m (usually several kilometers)

Interestingly, some field as well as laboratory studies questioned whether the "scale effect" exists. For example, Taylor and Howard (1987) concluded based on their study in a sandy aquifer that a distance-dependent dispersivity was not observed. Based on column experiments, Khan and Jury (1990) found that the dispersivity increased with increasing column length for undisturbed soil columns but was not dependent on length for repacked columns. Based on these studies one may conclude that scale effects may exist for some systems but not for others.

4.1 Definition of Scale Effects

Based on the above discussion, it is conceivable that the so-called scale effect has two meanings: one refers to the dispersivity (α) as a function of mean travel distance ($\bar{x}$) of a tracer; the other is a function of distance (x) from the source of a tracer solute. When plotted against a test scale, with scale being either mean travel distance or distance from a source, α increases with $\bar{x}$ or x or both. Specifically, the relation between α and mean travel distance $\bar{x}$ is obtained by fitting solute concentration profiles at different times with appropriate analytical or numerical solution of the CDE. Conversely, the relation between α and distance x is obtained from column studies by fitting breakthrough curves sampled at different distances or depths x. Therefore, it appears that the concept of scale effect is not well defined. If a clear definition of the concept of scale effect is not realized, one expects to encounter misunderstanding and misinterpretation of results. For example, one recognizes that, in a geologic system, if α increases with $\bar{x}$, then there exists a scale effect for such a system. Likewise, if α increases with distance x from source, we also recognize that there exists a scale effect. On the other hand, if we only know that a scale effect exists for a geological formation, we cannot distinguish beforehand whether α increases with mean travel distance $\bar{x}$ or with distance from source x. This situation is awkward, and the ambiguous meaning of the scale effects makes it difficult to implement. Suppose that we have sufficient data such that we can examine the relationship between α and $\bar{x}$ as well as that between α and x for the same geologic system. If it happens that α increases with $\bar{x}$ but not with distance x or α increases with distance x but not with mean travel distance $\bar{x}$, can one conclude that the system has a scale effect? Or, can one ignore that the system has a scale effect because α does not increase with distance from source x? We are not able to answer these questions because there is no unique concept of the scale effects.

4.1.1 Mean Travel Distance and Distance from Source

The above-mentioned confusion stems from the ambiguous meaning of the term "scale." When one refers to scale, one generally means the space scale such as the length of a soil column, the dimension (length and width) of an aquifer in a field transport experiment, or the area of the aquifer covered by a monitoring instrument. It should be pointed out that scale is a physically measurable quantitative property. In other words, a scale is a characteristic *index* associated with transport processes. For laboratory experiments, a scale could be the length of the soil column used. For field transport experiments, the potential scale could thus be infinite. Analogous to laboratory experiments, under certain circumstances, the distance between an observation well and an injection well could also be taken as a scale. In general, a space coordinate could be treated as a scale as long as the origin is set at the inlet of the soil column or the injection well. Obviously, scale is associated with distance from a source or a space coordinate and has nothing to do with time. In other words, scale should depend only on distance but not on time. Therefore, it is not appropriate to define scale in terms of $\bar{x}$. Mean travel distance $\bar{x}$ is actually not a distance from source or a space coordinate in the physical sense. On the contrary, $\bar{x}$ is a function of time or an expression for time. If $\bar{x}$ is taken as a scale, we encounter incorrectly the situation that scale varies with time. Like a space coordinate, scale is an independent variable and should not depend on any other variables. In laboratory experiments, this concept is straightforward and easily understood because a soil column length is the obvious parameter that can be associated with scale. However, in field experiments, we have difficulties in using this concept. In fact, several researchers use the mean travel distance $\bar{x}$ as a scale indiscriminately. Mean travel distance is theoretically where the solute front is at a certain time t. If one considers $\bar{x}$ as a scale, it turns out that the scale for a field experiment does not exist prior to the application of a solute pulse. Besides, because of the linear increase of $\bar{x}$ with time, scale is also a linear function of time in this case. These two deductions definitely do not sound correct. As mentioned above, scale is a characteristic parameter for a field site and is determined for a given system, for example, an experimental setup, regardless whether a transport experiment is actually conducted or when a transport experiment is initiated. If we have observation wells at different distances from an injection well, potentially we can say that this experiment is monitored at different scales. If the scale is defined based on $\bar{x}$, one is unable to predetermine the scale without prior knowledge of the schedule of the experimental sampling scheme. What $\bar{x}$ tells us is how far a solute front advances after release. So the validity of determining a scale based solely on $\bar{x}$ is questionable.

The term scale is frequently misrepresented because it often refers to the mean travel distance as well as the distance from the source. However, this is somewhat misleading and causes confusion. One source of confusion is that both $\bar{x}$ and x are used interchangeably. In other words, $\bar{x}$ can be replaced by x and vise versa. This type of interchange occurs quite often in the literature. For example, Equation [16] in Wheatcraft and Tyler (1988) is a relationship between α and $\bar{x}$. However, when this relationship was cited by Su (1995), it is converted to a relationship between α and x (see also Yates, 1990, 1992). We recognize that a relationship between dispersivity α and the space coordinate x may exist. However, we want to emphasize here that one cannot derive a relationship between α and x on the basis of a relationship between α and $\bar{x}$.

Another example of confusion is the reconstruction of the variance of travel distance σ_x^2 based on dispersivity α at a distance x rather than $\bar{x}$. Pickens and Grisak (1981b) reconstructed the variance-mean travel distance relationship based on dispersivity values measured at different distances by Peaudecerf and Sauty (1978). Sudicky and Cherry (1979) plotted reconstructed variances based on dispersivity from BTCs at different distances and those estimated from snapshots at different times in one graph to discuss the scale effects. The reconstruction of variance can be described as follows. Based on the assumption of homogeneous media, the variance of travel distance increases linearly with time such that

$$\sigma_x^2 = 2Dt \qquad (4.1)$$

where D is the dispersion coefficient. D can be expressed as:

$$D = \alpha v \qquad (4.2)$$

where v is the pore water velocity. We thus have

$$\sigma_x^2 = 2\alpha\bar{x} \qquad (4.3)$$

The relationship given by Equation 4.3 is often used to estimate α given both $\bar{x}$ and σ_x^2. However, we found that this relationship is also employed to reconstruct σ_x^2 (Pickens and Grisak, 1981b; Sudicky and Cherry, 1979). A dispersivity α measured at distance x is substituted into Equation (4.3). Therefore, the actual equation implemented reads:

$$\sigma_x^2 = 2\alpha x \qquad (4.4)$$

Thus, implicitly, an interchange from $\bar{x}$ to x is carried out. It should be pointed out that σ_x^2 is related only to $\bar{x}$ and not x. In other words, for a given time t or mean travel distance $\bar{x}$, one can compute σ_x^2 given α is known. However, no variance of travel distance σ_x^2 exists for a given distance x. The reason is that one needs a set or a collection of points to estimate a variance.

On the contrary, a variance of travel time (σ_t^2) exists with respect to a distance x. For example, one can estimate σ_t^2 from a BTC as follows:

$$\sigma_t^2 = \frac{\int_0^\infty (t - \bar{t})^2 c(t)dt}{\int_0^\infty c(t)dt}$$

where $c(t)$ is the solute concentration change over time at a fixed distance x (BTC), and $\bar{t}$ is the mean travel time and can be estimated as follows:

$$\bar{t} = \frac{\int_0^\infty tc(t)dt}{\int_0^\infty c(t)dt}$$

Therefore, σ_x^2 is associated with time (t) or mean travel distance ($\bar{x}$), whereas the variance of travel time σ_t^2 is associated with distance (x).

A linear relationship between σ_x^2 and t or $\bar{x}$ implies that a constant α for all times up to the maximum time under consideration could be estimated based on experimentally measured solute concentration profiles. Such an estimate of α applies to all distances from the source. Conversely, a dispersivity α estimated at a certain distance based on BTCs means a constant dispersivity is needed to described these BTCs at that specific distance for all times under consideration. If one should express the estimated dispersivity from a BTC at a certain distance in the variance-time (mean travel distance) format, it should be a straight line with a slope of 2α. Using Equation (4.3) to reconstruct σ_x^2 simply means that the obtained dispersivity α only applies to that specific $\bar{x}$ or time t. Therefore, to reconstruct variance based on α and distance x at which the dispersivity is estimated violates the assumption based on which the dispersivity is obtained. Clearly the examples of confusion described above stem from the ambiguous definition of scale.

Dispersivities measured at different distances from a given source are often compared with dispersivities measured at different $\bar{x}$ or time t (Pickens and Grisak, 1981a; Arya et al., 1988; Gelhar, Welty, and Rehfeldt, 1992; Neuman, 1990). In addition, variation in dispersivities is almost always attributed to the scale under which it is estimated. The heterogeneity or type of formation of the geological systems is often ignored. The other problem is that the integrity of transport processes is often ignored when a regression model is applied to dispersivities estimated based on different transport processes and from different media.

Dispersivity has been estimated in both laboratory and field studies because of its importance to the governing CDE. A comparison is often made among measured dispersivity values for different scales to support the finding that dispersivity is scale dependent. No matter whether the dispersivities were estimated for a distance from a source or for a certain time in terms of mean travel distance, they were compared in terms of a quantitative index: scale (Gelhar, Welty, and Rehfeldt, 1992; Neuman, 1990). As discussed above, the definition of the term scale is not clear. Therefore, whether this comparison is reasonable remains open. Although both distance from a source x and mean travel distance $\bar{x}$ share a common dimension of length, these two terms are quite different according to their physical meanings. Conceptually, dispersivities obtained under different mean travel distance and different distance are not comparable in terms of length scale, which is the quantitative value of either x or $\bar{x}$. The reason is that the underlying assumptions for the estimation of dispersivity with respect to distance and those with respect to mean travel distance are actually exclusive to each other as discussed above.

When comparisons among dispersivities are made, values corresponding to different scales are often from different media, that is, porous media and fractures, instead of the same medium. This is understandable because of difficulties in obtaining dispersivity information from the same medium with scales varying from tens of centimeters to hundreds of meters. Beyond the comparison of dispersivity is the regression of dispersivity versus scale. Based on dispersivity data available in the literature, several studies have been conducted to develop a functional relationship between dispersivity values and associated scale based on regression analysis (Arya et al., 1988; Neuman, 1990; Xu and Eckstein, 1995). Furthermore, a universal dispersivity value for a specific scale is determined based on the function obtained. Mathematically, this kind of regression analysis is feasible. However, rigorous theoretical development is needed in support of the regression analysis. Currently, several models are available to describe the relationship between α and $\bar{x}$. For example, Zhou and Selim (2002) developed a model to describe time-dependent dispersivity. In addition, others developed stochastic models that describe dispersivity-mean travel distance based on heterogeneity of the hydraulic conductivity in porous media (Neuman and Zhang, 1990; Zhang and Neuman, 1990). These models are commonly used to support regression of measured α versus different scales representing several distances and mean travel distances over different sites (Neuman, 1990; Wheatcraft and Tyler, 1988). To justify the regression of data from different sites, however, a separate theory must be developed. In fact, both fractal and stochastic models describe an instantaneous relationship between dispersivity and time or mean travel distance. Dispersivity takes a distinct value for a specific time. The measured α available in the literature, however, is an apparent or average α over the period up to the time at which the dispersivity is estimated. Thus, the estimated dispersivity under such condition is not actually applicable to the models discussed above.

Discrepancies regarding use of the terms scale and scale effects are mainly caused by the ambiguous definition of the term scale. In fact, we are dealing with four types of relations between dispersivity and time and/or distance rather than one universal dispersivity-scale relationship.

1. The first is a time-averaged dispersivity versus time (or mean travel distance). Here apparent dispersivity is estimated at different discrete times or mean travel distances. When plotted, dispersivities

may increase with time as reported by several studies (Freyberg, 1986; Sudicky, Cherry, and Frind, 1983). Though dispersivity is estimated according to a snapshot at a certain time t, such α represents an average value for the time period up to the time t.

2. The second is a time-dependent dispersivity as described by fractal and stochastic models (Zhou and Selim, 2002; Neuman, 1990; Arya, 1988). This relationship reveals that α is a continuous function of time. For each time t, one has an instantaneous value of α for that specific time.

3. The third is α as a continuous function with distance from source. This type of relationship is not actually observed experimentally or developed theoretically. Rather it is derived by replacing the mean travel distance $\bar{x}$ with x distance from source in the second case (Su, 1995; Yates, 1990, 1992). Removal of the bar from mean travel distance $(\bar{x})$ changes the relationship to a dispersivity-distance relationship. One may argue that the first type of relationship is a description of dispersivity versus scale and this derived relationship should hold in terms of scale (distance from source is also a scale). As discussed above, the distance from the source and mean travel distance are not interchangeable. If dispersivity can vary with distance from the source according to such a trend, theoretical support is needed to support such an expression.

4. The fourth type of dispersivity versus scale is a distance-averaged α versus the length scale in consideration (Zhang, Huang, and Xiang, 1994; Butters and Jury, 1989; Burns, 1996; Khan and Jury, 1990; Pang and Close, 1999). In this case, dispersivity for different column lengths or distances between observation wells and an injection well is obtained through fitting of observed BTCs. In some cases, dispersivity values were found to increase with the length scale under consideration, that is, the column length or the distance between the observation and injection wells. What such a relationship means is that a constant dispersivity for the whole column is needed to predict the breakthrough process. However, for each column length, a different (constant) dispersivity must be used. Thus, one cannot conclude from this observation that dispersivity is distance dependent. Actually, in the fourth case, for each individual length, the basic assumption is that the dispersivity is constant along the whole column up to the length considered. The dispersivity value could thus be considered as an average or apparent value.

Rather than one so-called generic dispersivity-scale relationship, we emphasize here four distinct dispersivity-time or dispersivity-distance relationships. Among these four relationships, only the fourth could possibly be characterized by the so-called dispersivity-scale relationship if one defines scale as length of column or distance from source.

4.2 Case Studies

As pointed out earlier, distance-dependent dispersivity has been used rather than time-dependent dispersivity in terms of dispersivity-mean travel distance relationship. In the previous section, we emphasized the differences between mean travel distance $\bar{x}$ and distance from source x. Such differences have been often ignored. Given an expression for dispersivity-mean travel distance, for example, $\alpha = 0.1\bar{x}$, we investigated the effect on BTCs and solute concentration profiles if the bar from $\bar{x}$ is removed and thus yields a dispersivity-distance relationship, that is, $\alpha = 0.1x$. We used the finite-difference approach to solve the CDE with a time-dependent (in terms of $\bar{x}$) or distance-dependent dispersivity. We focused on the differences in BTCs and concentration profiles resulting from different dispersivity models. We also propose a procedure for obtaining a distance-dependent dispersivity model.

To compare the differences in transport processes in media with time-dependent dispersivity and distance-dependent dispersivity, we have to solve the CDE with time-dependent dispersivity or distance-dependent dispersivity. An analytical solution is available for several dispersivity-distance models. Yates (1990) suggested an analytical solution for one-dimensional transport in heterogeneous porous media with a linear distance-dependent dispersion function. Su (1995) developed a similar solution for media with a linearly distance-dependent dispersivity. Yates (1992) gave an analytical solution for one-dimensional transport in porous media with an exponential dispersion function. Logan (1996) extended the work of Yates (1990, 1992) and developed an analytical solution for transport in porous media with an exponential dispersion function and decay. Transport in porous media with time-dependent dispersion function has been studied (Zou, Xia, and Koussi, 1996). Generally, dispersivity is expressed directly as a function of time t instead of mean travel distance $\bar{x}$. As far as we know, analytical solutions for transport in porous media with time-dependent dispersion function in terms of $\bar{x}$ are not available. Although analytical solutions for some cases are available, they were not often used because they are difficult to use. Besides, numerical evaluation is often necessary to compute the analytical solution. In this investigation, we solved the CDE with time-dependent dispersivity or distance-dependent dispersivity using numerical methods. As an illustration, we chose the linear or power law form as a model for time-dependent dispersivity. The advantage of using a power law model lies in that it will either recover a linear model or reduce to a constant (homogeneous case) if we set the exponent term to proper values.

4.2.1 Generalized Equations and Simulations

4.2.1.1 CDE with a Linearly Time-Dependent Dispersivity

For a linearly time-dependent dispersivity in terms of mean travel distance $\bar{x}$, a representative model can be expressed as (Pickens and Grisak, 1981b):

$$\alpha(\bar{x}) = a_1 \bar{x} \tag{4.5}$$

where a_1 is a dimensionless constant. Physically, a_1 is the slope of dispersivity-mean travel distance line. If we ignore molecular diffusion, the dispersion coefficient can be written as:

$$D(t) = \alpha(\bar{x})v = a_1 \bar{x} v = a_1 v^2 t \tag{4.6}$$

where v is mean pore water velocity. Accordingly, the governing equation in a heterogeneous system with a time-dependent dispersion coefficient is given by:

$$\frac{\partial c}{\partial t} = \frac{\partial}{\partial x}\left(D(t) \frac{\partial c}{\partial x}\right) - v \frac{\partial c}{\partial x} \tag{4.7}$$

Clearly, the dispersion coefficient as determined by Equation (4.6) is constant for the entire domain at any fixed time. Therefore, the governing equation can be rewritten as:

$$\frac{\partial c}{\partial t} = a_1 v^2 t \frac{\partial^2 c}{\partial x^2} - v \frac{\partial c}{\partial x} \tag{4.8}$$

In addition, the appropriate initial and boundary conditions for a finitely long soil column can be expressed as:

$$c(x,t) = 0, \quad t = 0 \tag{4.9}$$

$$vc(x,t) = vc_0 - a_1 v^2 t \frac{\partial c}{\partial x}, \quad x = 0, \quad 0 < t \leq T \tag{4.10}$$

$$vc(x,t) = -a_1 v^2 t \frac{\partial c}{\partial x}, \quad x = 0, \quad t > T \tag{4.11}$$

$$\frac{\partial c}{\partial x}\bigg|_{x = L} = 0 \tag{4.12}$$

where L is the length of the soil column, T is the input pulse duration, and c_0 is the solute concentration in the input pulse. The governing equation (4.8) subject to initial and boundary conditions (Equations 4.9 through

4.12) can be solved using finite-difference methods. The detailed finite-difference scheme is shown in Appendix 4A. The resulting tri-diagonal linear equation system was solved using the Thomas algorithm (Press et al., 1992).

4.2.1.2 CDE with a Linearly Distance-Dependent Dispersivity

By the removal of the bar from the mean travel distance ($\bar{x}$) in the dispersivity-mean travel distance relationship, we change the mean travel distance to a distance from source (x) and obtain a distance-dependent dispersivity as given by

$$\alpha(x) = a_2 x \tag{4.13}$$

where now a_2 is a constant as a_1 is in Equation 4.5. If we also ignore molecular diffusion, the dispersion coefficient becomes a function of distance from source and is thus given by

$$D(x) = \alpha(x)v = a_2 x v \tag{4.14}$$

Therefore, transport for a tracer solute or nonreactive chemical in a one-dimensional heterogeneous soil system with distance-dependent dispersion coefficient, under steady-state water flow, is governed by the following equation:

$$\frac{\partial c}{\partial t} = \frac{\partial}{\partial x}\left(D(x)\frac{\partial c}{\partial x}\right) - v\frac{\partial c}{\partial x} \tag{4.15}$$

Substituting Equation 4.14 into the above governing equation and expanding gives:

$$\frac{\partial c}{\partial t} = a_2 v x \frac{\partial^2 c}{\partial x^2} - (1 - a_2)v\frac{\partial c}{\partial x} \tag{4.16}$$

The corresponding initial and boundary conditions for a finite soil column can be expressed as:

$$c(x,t) = 0, \quad t = 0 \tag{4.17}$$

$$c(x,t) = c_0, \quad x = 0, \quad 0 < t \leq T \tag{4.18}$$

$$c(x,t) = 0, \quad x = 0, \quad t > T \tag{4.19}$$

$$\left.\frac{\partial c}{\partial x}\right|_{x=L} = 0 \tag{4.20}$$

where L, T, and c_0 are the same as in Equations 4.10 through 4.12. The third-type boundary condition is applied to the upper boundary. However, because the dispersion coefficient vanishes at $x = 0$, the third-type boundary condition formally reduces to the first-type boundary condition. The governing equation (4.16) subject to initial and boundary conditions (Equations 4.17 through 4.20) were solved numerically (see Appendix 4B for the detailed finite-difference scheme).

4.2.1.3 CDE with a Nonlinearly Time-Dependent Dispersivity

Zhou and Selim (2002) developed a fractal model to describe a time-dependent dispersivity in terms of mean travel distance $\bar{x}$. The fractal model reads:

$$\alpha(\bar{x}) = a_3 \bar{x}^{D_{fr}-1} \tag{4.21}$$

where now a_3 is a constant with dimension $L^{2-D_{fr}}$, and D_{fr} is the fractal dimension of the tortuous stream tubes in the media. D_{fr} varies from 1 to 2. If $D_{fr} = 1$, we recover the time-invariant constant dispersivity. Similarly, if $D_{fr} = 2$, Equation 4.21 reduces to Equation 4.5. Again, we assume molecular diffusion can be ignored and dispersion coefficient for a nonlinear dispersion function is given by

$$D(t) = \alpha(\bar{x})v = a_3 \bar{x}^{D_{fr}-1}v = a_3 v^{D_{fr}} t^{D_{fr}-1} \tag{4.22}$$

Substituting Equation 4.22 into Equation 4.7 and rearranging yields the following governing equation:

$$\frac{\partial c}{\partial t} = -v\frac{\partial c}{\partial x} + a_3 v^{D_{fr}} t^{D_{fr}-1}\frac{\partial^2 c}{\partial x^2} \tag{4.23}$$

Equation 4.8 is recovered if we let $D_{fr} = 2$ in the above equation. The upper boundary conditions are

$$vc(x,t) = vc_0 - a_3 v^{D_{fr}} t^{D_{fr}-1}\frac{\partial c}{\partial x}, \quad x = 0, \quad 0 < t \leq T \tag{4.24}$$

$$vc(x,t) = -a_3 v^{D_{fr}} t^{D_{fr}-1}\frac{\partial c}{\partial x}, \quad x = 0, \quad t > T \tag{4.25}$$

The remaining initial and lower boundary conditions are the same as those for linear dispersivity model (Equations 4.9 and 4.12). The above system was also solved using the finite difference method (see Appendix 4C).

4.2.1.4 CDE with a Nonlinearly Distance-Dependent Dispersivity

If we remove the bar from $\bar{x}$ in Equation 4.21, we obtain the following non-linearly distance-dependent dispersivity:

$$\alpha(x) = a_4 x^{D_{fr}-1} \qquad (4.26)$$

where now a_4 is a constant with dimension $L^{2-D_{fr}}$. Under this condition, the dispersion coefficient D is given by:

$$D(x) = \alpha(x)v = a_4 x^{D_{fr}-1}v \qquad (4.27)$$

Accordingly, the governing equation now reads:

$$\frac{\partial c}{\partial t} = a_4 v x^{D_{fr}-1}\frac{\partial^2 c}{\partial x^2} - \left[1 - a_4(D_{fr}-1)x^{D_{fr}-1}\right]v\frac{\partial c}{\partial x} \qquad (4.28)$$

Equation 4.28 subject to initial and boundary conditions (Equations 4.17 through 4.20) was solved with the finite-difference method (see Appendix 4D).

4.2.1.5 Comparison of Models and Simulations

For the CDE with a time-dependent dispersivity, the magnitude of the dispersivity α increases with mean travel distance or time. Under this situation, the dispersivity value remains constant over the entire spatial domain. In other words, the entire medium is treated as a homogeneous system with a fixed constant dispersivity value for each specific time. On the contrary, if one removes the bar from the mean travel distance in the dispersivity-mean travel distance relationship, the dispersive property of the medium is completely altered. As a result, dispersivity becomes a function of distance from source instead of mean travel distance or time. Under such conditions, the dispersivity is held constant over the time being considered for any location but increases with distance from the source where the solute is released. Therefore, the resulting parameter fields and thus the governing equations are quite different and depend on whether the bar in mean travel distance is removed. An extra term occurs for the distance-dependent dispersivity models in the governing equation to account for the dependency of dispersivity on distance.

Differences in the governing equations inevitably induce the differences in the numerical scheme (finite difference equations, see the Appendices for details). The dependence of dispersivity on time or distance is carried over to the finite-difference approximations. Corresponding to time dependence, index j, which indicates time domain discretization, occurs in the finite-difference approximation for the governing equation with dispersivity as a function of mean travel distance. Accordingly, index i appears for a distance-dependent dispersivity. Comparison between the finite-difference equations only is not enough to confirm the difference between the two processes described by these two different governing equations. One needs to show differences in BTCs as well as solute distribution along spatial coordinate or flow direction to achieve a generalized conclusion. For comparison, both governing equations subject to the same initial and boundary conditions were considered here. Because of the complexity of the difference equations, it is difficult to assess convergence conditions. The time and space increments were based on the governing equation with a constant dispersivity, which in our case equals the coefficient *a* in the dispersivity function. The assessment of the convergence of numerical approximation was achieved through mass balance calculations as well as the magnitude and oscillation of the resulting numerical solutions for solute concentration.

The parameters used for our simulations are given in Table 4.1. Similar parameter values were selected for both cases. The only difference lies in that the variable mean travel distance for the time-dependent dispersivity is replaced with distance from source to generate the distance-dependent dispersivity. Two different column lengths were considered. One is 50 cm in length, the other 100 cm. A longer pulse length is used for the 100-cm column to obtain comparable BTCs. BTCs as well as distribution profiles were compared for a solute tracer under different scenarios.

Simulated BTCs with linearly time-dependent and distance-dependent α are shown in Figure 4.1 for 50- and 100-cm soil columns. The BTCs from either time-dependent or distance-dependent α appear somewhat similar. Nevertheless, several distinct features are apparent. Based on our simulations, the column length showed modest influence in the relative relationship between BTCs of the media with a time-dependent dispersivity and those with a distance-dependent dispersivity. For both long and short columns, distance-dependent dispersivity resulted in earlier arrival of the BTC than the time-dependent counterpart. Both BTCs exhibited similar leading

TABLE 4.1

Parameters Used for Simulations of Time-Dependent and Distance-Dependent Dispersivities (α)

Parameter	Time-Dependent Dispersivity	Distance-Dependent Dispersivity
Moisture content (cm^3/cm^3)	0.40	0.40
Column length (cm) (short/long)	50.0/100.0	50.0/100.0
Water flux rate (cm/h)	5.0	5.0
Initial concentration (mg/L)	0.0	0.0
Concentration in input pulse (mg/L)	10.0	10.0
Pulse duration (hour) ($L = 50$ cm/100 cm)	2.0/16.0	2.0/16.0
Dispersivity α (cm)	$0.5\bar{x}$	$0.5x$

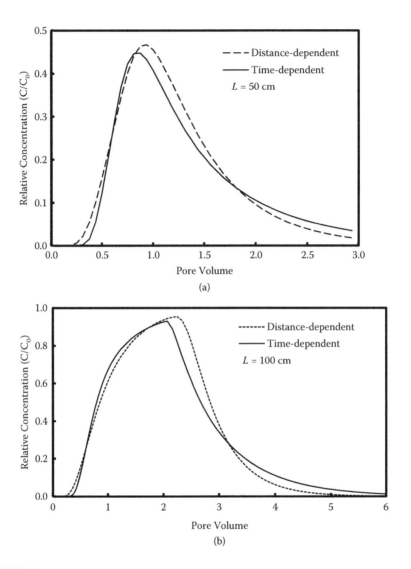

FIGURE 4.1
Comparison of simulated breakthrough curves based on time-dependent (Equation 4.8) and distance-dependent (Equaton 4.16) dispersivity for 50 cm (top) and 100 cm (bottom) columns.

edge, however. BTCs for the distance-dependent cases showed higher peak concentrations than those for the time-dependent cases. In general, time-dependent α resulted in enhanced tailing compared with distance-dependent counterparts.

Snapshots for solute distributions at different times for the 100-cm column with linear time-dependent or distance-dependent α are shown in Figure 4.2. Because the dispersion coefficient vanishes at the inlet ($x = 0$) for

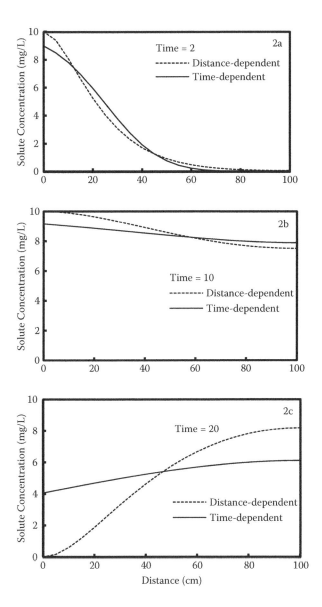

FIGURE 4.2

Comparison of simulated solute concentration profiles of a pulse tracer based on time-dependent (Equation 4.8) and distance-dependent (Equation 4.16) dispersivity after 2, 10, and 20 hours of transport. Column length $L = 100$ cm and pulse duration = 16 h.

the distance-dependent α, the concentration at the inlet is always at c_0 for all times during pulse application (16 h). At early times, solute distribution profiles appear similar for both types of α models. However, the two snapshots separated gradually over time. Generally, solute concentration profiles exhibited a rapid decrease or sharp fronts when distance-dependent α was used. In other words, distance-dependent α resulted in a steeper solute concentration profile than time-dependent α. At later times, the solute fronts advanced further in the porous media when distance-dependent α was used. Our results clearly demonstrate the differences between transport processes in a medium with time-dependent and that with distance-dependent α.

4.3 Effects of Nonlinearity of the Dispersivity Model

The nonlinear dispersivity models (Equations 4.21 through 4.28) provide an opportunity to investigate the effects of fractal dimension on solute BTCs. For computational convenience, only 10-cm-long soil columns were considered in this section (Table 4.2). For time-dependent dispersivity, the BTCs for different fractal dimension D_{fr} are compared in Figure 4.3. From this figure, we can see that the fractal dimension has significant influences on the overall shape of the BTCs. As D_{fr} increases, the initial arrival time becomes shorter with lower peak concentrations. Moreover, the BTCs exhibited enhanced tailing or increased spreading as D_{fr} value approached 2.0.

Comparison of BTCs for distance-dependent α having different exponent values is shown in Figure 4.4. For distance-dependent dispersivity, higher values of D_{fr} resulted in earlier arrival of BTC, a lower peak concentration, and an enhanced tailing in the BTCs. However, our simulations clearly show that the differences among BTCs for different D_{fr} shown in Figure 4.4 are

TABLE 4.2

Parameters Used in Simulation to Examine the Effect of Exponents in the Dispersivity Expressions of Equations 4.21 and 4.26

Parameter	Time-Dependent α	Distance-Dependent α
Moisture content (cm³/cm³)	0.40	0.40
Column length (cm)	10.0	10.0
Water flux rate (cm/h)	5.0	5.0
Initial concentration (mg/L)	0.0	0.0
Concentration in input pulse (mg/L)	10.0	10.0
Pulse duration (h)	2.0	2.0
Dispersivity α (cm)	$0.5\bar{x}^{D_{fr}-1}$, $D_{fr} = 1.25, 1.50, 1.75, 2.0$	$0.5x^{D_{fr}-1}$, $D_{fr} = 1.25, 1.50, 1.75, 2.0$

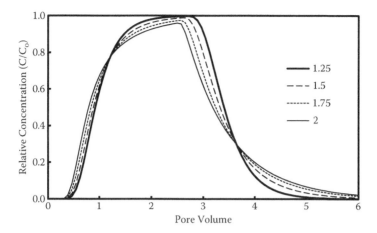

FIGURE 4.3
Comparison of simulated breakthrough curves based on time-dependent dispersivity (Equation 4.23) with different fractal dimensions D_{fr}.

relatively small compared with the time-dependent counterpart shown in Figure 4.3.

4.3.1 Transfer Function Models

The CDE is the most commonly used model to describe solute movement through soils where perfect lateral mixing is assumed. Thus, the

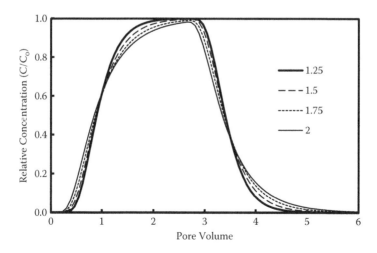

FIGURE 4.4
Comparison of simulated breakthrough curves based on distance-dependent dispersivity (Equation 28) with different exponents D_{fr}.

CDE predicts a linear increase of travel time variance with distance. With an approach similar to that used by Mercado (1967), Jury and Roth (1990) developed a stochastic-convective model. The stochastic-convective process assumes that the solute moves at different velocities in isolated stream tubes without lateral mixing. Use of log normal distribution of travel time results in a convective log normal transfer function model (CLT). The CLT model describes solute transport characterized by a quadratic increase in the travel time variance with depth. However, the travel time variance is often reported to increase nonlinearly with distance because of heterogeneity of the media (Zhang, Huang, and Xiang, 1994). To account for the nonlinearity in the relationship between travel time variance and distance or depth, Liu and Dane (1996) proposed an extended transfer function model (ETFM). They introduced an additional parameter to represent the degree of lateral solute mixing. The ETFM can be considered as a transition between the CDE and the CLT. Zhang (2000) also proposed an extended convective log normal transfer function model (ECLT). Meanwhile, he attempted to unify all types of transfer function model with a generalized transfer function model (GTF). More important, Zhang (2000) showed that the distance-dependent dispersivity model could be associated with the parameters of the GTF.

The mean and variance of the travel time are two important characteristic elements in transfer function theory. A third parameter related to scale effects is the coefficient of variation (CV) of the travel time. The squared CV at depth z is given by:

$$CV(t,z)^2 = \frac{\mathrm{Var}(t,z)}{E(t,z)^2} \qquad (4.29)$$

where $\mathrm{Var}(t,z)$ and $E(t,z)$ are the variance and mean of the travel time at depth z. For the CDE, the means, variances, and CVs of the travel time at depth z and l are related by:

$$\frac{E(t,z)}{E(t,l)} = \frac{z}{l}\frac{\mathrm{Var}(t,z)}{\mathrm{Var}(t,l)} = \frac{z}{l}\frac{CV(t,z)^2}{CV(t,l)^2} = \frac{l}{z} \qquad (4.30)$$

For the CLT, similar relationships can be established as:

$$\frac{E(t,z)}{E(t,l)} = \frac{z}{l}\frac{\mathrm{Var}(t,z)}{\mathrm{Var}(t,l)} = \left(\frac{z}{l}\right)^2\frac{CV(t,z)^2}{CV(t,l)^2} = 1 \qquad (4.31)$$

For the ECLT (Zhang, 2000), the relationships are

$$\frac{E(t,z)}{E(t,l)} = \left(\frac{z}{l}\right)^{\lambda}\frac{\mathrm{Var}(t,z)}{\mathrm{Var}(t,l)} = \left(\frac{z}{l}\right)^{2\lambda}\frac{CV(t,z)^2}{CV(t,l)^2} = 1 \qquad (4.32)$$

The exponent λ is introduced by Zhang (2000) to describe transport processes in which the travel time of solute may increase with depth nonlinearly. For the ETFM by Liu and Dane (1996), the relationships are

$$\frac{E(t,z)}{E(t,l)} = \frac{z}{l}\frac{\mathrm{Var}(t,z)}{\mathrm{Var}(t,l)} = \left(\frac{z}{l}\right)^{2a}\frac{CV(t,z)^2}{CV(t,l)^2} = \left(\frac{l}{z}\right)^{2(1-a)} \tag{4.33}$$

The value of parameter a in the above equation lies in the range between 0.5 and 1 ($0.5 \leq a \leq 1$). Based on the observation of the above relationships for different transfer function models, Zhang (2000) proposed a generalized relationship of means, variances, and CVs of the travel time at depth z and l for a GTF:

$$\frac{E(t,z)}{E(t,l)} = \left(\frac{z}{l}\right)^{\lambda_1}\frac{\mathrm{Var}(t,z)}{\mathrm{Var}(t,l)} = \left(\frac{z}{l}\right)^{\lambda_2}\frac{CV(t,z)^2}{CV(t,l)^2} = \left(\frac{l}{z}\right)^{2(\lambda_1-\lambda_2)} \tag{4.34}$$

where λ_1 and λ_2 are parameters of the time moments.

The dispersivity can be estimated based on the CV at depth z and is given by:

$$\alpha = \frac{z}{2}CV(t,z)^2 \tag{4.35}$$

Substituting the CV for the GTF (Equation 4.34) into the above equation gives:

$$\alpha \propto z^{1+2(\lambda_2-\lambda_1)} \tag{4.36}$$

If the two parameters λ_1 and λ_2 in Equation 4.36 satisfy $\lambda_2 - \lambda_1 = -0.5$, dispersivity is constant with the distance (CDE). Otherwise, Equation 4.36 describes a distance-dependent dispersivity.

4.3.2 Fractional CDE in Porous Media

An alternative to the scale approach discussed above is that of the fractional convection-dispersion equation (FCDE) to describe anomalous transport phenomena in aquifers (Benson et al. 2000a). Cushman and Ginn (2000) showed that the FCDE is a special case of the convolution-Fickian nonlocal advection-dispersion equation (ADE) proposed earlier by Cushman and Ginn (1993).

The one-dimensional symmetrical FCDE reads:

$$\frac{\partial c}{\partial t} = -v\frac{\partial c}{\partial x} + \frac{1}{2}D\left[\frac{\partial^\alpha c}{\partial x^\alpha} + \frac{\partial^\alpha c}{\partial(-x)^\alpha}\right] \tag{4.37}$$

where D is the fractional dispersion coefficient (L^aT^{-1}) and the superscript a is the order of fractional differentiation, $0 < \alpha \leq 2$. It is worth noting that the classic CDE is recovered when the fractional order of FCDE is set to 2. Fractional derivatives are integro-differential operators defined as (Podlubny, 1999):

$$\frac{\partial^\alpha c}{\partial x^\alpha} = \frac{1}{\Gamma(k-\alpha)}\left(\frac{\partial^k}{\partial x^k}\right)\int_{-\infty}^{x}(x-\xi)^{k-\alpha-1}c(\xi,t)d\xi \qquad (4.38)$$

$$\frac{\partial^\alpha c}{\partial(-x)^\alpha} = \frac{(-1)^k}{\Gamma(k-\alpha)}\left(\frac{\partial^k}{\partial x^k}\right)\int_{x}^{\infty}(\xi-x)^{k-\alpha-1}c(\xi,t)d\xi \qquad (4.39)$$

where $\alpha > 0$, Γ is the gamma function, and k is the smallest integer number larger than α. If the value of α is a whole number, fractional derivatives reduce to ordinary derivatives. Some properties of fractional derivative are given in the appendix of Benson et al. (2000a).

Benson et al. (2000b) applied the FCDE to two cases: a laboratory sandbox transport experiment and the Cape Cod field-scale transport experiment. The fractional order for the laboratory data was found to be 1.55, whereas the Cape Cod bromide plumes could be modeled using an FCDE with an order of 1.65 to 1.8. Pachepsky, Benson, and Rawls (2000) simulated scale-dependent solute transport in soils using the. They also presented a comparison between the FCDE and ADE based on statistical analysis. They found that the FCDE could describe transport processes of a solute tracer better. Benson et al. (2000b) also provided a method to estimate the fractional order a and the fractional dispersion coefficient D separately. However, in our opinion, a simultaneous estimation of both α and D is more appropriate. Specific reasons will be discussed later in this chapter. Benson et al. (2000b) estimated the fractional order based on the relationship between measured apparent dispersivity and distance from the sandbox experiment. Pachepsky, Benson, and Rawls (2000) justified the application of the FCDE based on the scale-dependent transport phenomenon. Zhou and Selim (2003) argued that the order of FCDE α cannot be associated with scale-dependent transport directly. In fact, the relationship between α and scale-dependent dispersivity is not theoretically supported and application of the FCDE needs to be justified.

4.3.3 Comparison between CDE and FCDE

As indicated above, the classical ADE predicts a linear increase of the variance of travel distance with time or mean travel distance, whereas the FCDE predicts a nonlinear increase of the variance of travel distance. Therefore, comparing linear models of variance pattern with nonlinear ones is equivalent to direct comparison between ADE and FCDE. The comparison between these two models is achieved by conducting the appropriate F-test (Green and Caroll, 1978).

The relationship of the measured variance of travel distance to the mean travel distance or time (i.e., σ_x^2 versus $\bar{x}$ or σ_x^2 vs t) can thus be fitted using linear or nonlinear models through linear and nonlinear least-squares fitting techniques. Because of the simplicity of the linear least-squares method, nonlinear models are often converted to linear format by taking logarithm transformation. Therefore, there are three options to model the $\sigma_x^2 - \bar{x}$ or $\sigma_x^2 - t$ relationship:

1. Linear coordinate: the linear model (model I):

$$Y = A \times X \tag{4.40}$$

2. Linear coordinate: the nonlinear model (model II—nonlinear least-squares fitting):

$$Y = A \times X^B \tag{4.41}$$

3. Log-log coordinate: the nonlinear model (model III—linear least-squares fitting):

$$\log Y = \log 10^a + b \log X \tag{4.42}$$

where Y refers to the variance of travel distance σ_x^2, X is the mean travel distance $\bar{x}$ or time t, and A, B, 10^a, and b are regression coefficients that can be used to determine the transport parameters. Models II and III are equivalent. For a linear model, the regression coefficient A is twice the longitudinal dispersivity or dispersion coefficient, depending on whether X refers to the mean travel distance or time. For nonlinear models, the power B or b can be used to determine the fractional order a. Model I of Equation 4.37 represents the CDE equation where α equals 2.

Zhou and Selim (2003) carried out a comparison of the different models in an attempt to justify use of the FCDE. Specifically, a comparison between different models is to determine whether an FCDE is necessary and whether the FCDE is superior to the classical CDE in description of a given data set. In their investigation, the data from the Cape Cod site are used as an example. We fitted the above three models with the data from Garabedian et al. (1991). The functional relationships between σ_x^2 versus $\bar{x}$ as well as σ_x^2 versus t were obtained.

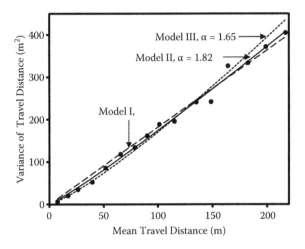

FIGURE 4.5

Comparison among different models: variance versus mean travel distance. Data from Cape Cod field experiments (Garabedian et al., 1991).

If the estimated α is not significantly different from 2, it is not necessary to use a fractional order other than 2. A way to test the significance of α is to check whether $\alpha = 2$ is included in the 95% confidence interval of the estimated α. Zhou and Selim (2003) obtained fractional order α values based on the relationship between variance and mean travel distance of 1.82 and 1.65 for model II and model III, respectively. Both values are significantly less than 2.0. However, if we estimate the fractional order based on the variance-time relationship, the values for α were 1.91 and 1.79 for model II and model III, respectively. Among the two values, only the second ($\alpha = 1.79$ based on model III) is significantly less than 2.0. These values for the fractional order are almost the same as those reported in Benson et al. (2000b). Therefore, different estimated values for a were obtained when the same data set was fitted using the two methods, that is, direct nonlinear fitting of the original data versus linear fitting of log-transformed data (see Figures 4.5 and 4.6).

Based on the above analysis, the use of the FCDE often needs justification because its solution is more complex than that of the classical CDE. If the order of an FCDE is not significantly smaller than 2 or no improvement in describing the transport process is achieved, use of the FCDE is unjustified. Nevertheless, the dispersivity-time or dispersivity-mean travel distance relationship is needed to estimate the fractional order α. The use of the dispersivity-distance relationship to estimate the fractional order is theoretically unjustified and may result in unreasonable α values.

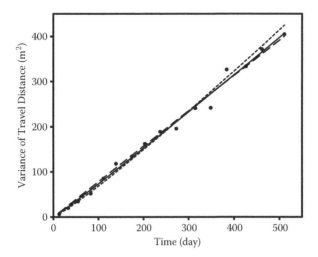

FIGURE 4.6
Comparison among different models: variance versus time. Data from Cape Cod field experiments (Garabedian et al., 1991).

4.4 Fractals

Another alternative to describe a time-dependent dispersivity in terms of mean travel distance x using fractal models has been proposed. Zhou and Selim (2003) proposed a modified fractal as follows. If one assumes a Fickian process occurs when a particle moves along the fractal tortuous streamline, the variance of travel distance at a certain time t is given by:

$$\sigma_x^2 = 2\alpha_0 \bar{x}_{fr} \tag{4.43}$$

where α_0 is a constant, time-independent true dispersivity, and $\bar{x}_{fr}$ is the mean path length of all possible fractal streamlines traveled by solute particles till time t. Taking expressions on both sides of Equation 4.14 and substituting it in the above equation yields:

$$\sigma_x^2 = 2\alpha_0 \varepsilon_c^{1-D} E[x^D] \tag{4.44}$$

where $E[x^D]$ stands for the mean value of x^D. In fact, $E[x^D]$ is the Dth-order moment of the travel distance. The difficulty here lies in that D is a fractional number rather than an integer. If we write x^D as $[\bar{x} - (\bar{x} - x)]^D$ and expand it in terms of a power series of $\bar{x}^{D-k}(\bar{x}-x)^k$ with $k = 0,1,2,\ldots$, then, we may reach the following expression for σ_x^2:

$$\sigma_x^2 = \frac{2\alpha_0\varepsilon_c^{1-D}\bar{x}^D}{1 - D(D-1)\alpha_0\varepsilon_c^{1-D}\bar{x}^{D-2}} \tag{4.45}$$

The above equation shows that the relationship between the variance of travel distance σ_x^2 and the mean travel distance $\bar{x}$ is rather complex if D is greater than 1. Approximately, σ_x^2 grows proportionally to $\bar{x}$ raised to the power of D. If D reduces to 1, the above equation reduces to the nonfractal case, that is, σ_x^2 grows linearly with $\bar{x}$. On the other hand, if D is equal to 2, Equation 4.45 becomes

$$\sigma_x^2 = \frac{2\alpha_0\bar{x}^2}{\varepsilon_c - 2\alpha_0} \tag{4.46}$$

Obviously, Equation 4.46 is consistent with the Mercado model (Pickens and Grisak, 1981a). Since σ_x^2 is always positive, the denominator of the right side of Equation 4.46 must be greater than zero. This restriction implies that the fractal cutoff limit, if $D = 2$, must satisfy the following:

$$\varepsilon_c > 2\alpha_0 \tag{4.47}$$

Based on fractal geometry, Equation 4.47 also casts restriction on the travel distance x. That is, the shortest apparent straight-line travel distance must at least have the length of the fractal cutoff limit in order to use it to represent the actual length of stream lines with a fractal dimension other than 1. If we assume the apparent travel distance follows a normal distribution, then the shortest apparent straight-line travel distance can be estimated by:

$$x_{\min} = \bar{x} - 3\sigma_x \tag{4.48}$$

where $x_{\min}$ stands for the shortest straight-line travel distance after time t with the mean travel distance given by $\bar{x}$. Similarly, the mean travel distance $\bar{x}$ in Equation 4.45 must be large enough so that the second term in the denominator of the right side of the equation is less than 1. That is, to use Equation 4.45 to show the time dependence of dispersivity, $\bar{x}$ must abide by:

$$\bar{x} > \frac{(2\alpha_0)^{1/(2-D)}}{\varepsilon_c^{(D-1)/(2-D)}} \tag{4.49}$$

where $1 < D < 2$. The above discussion shows that only after the travel distance and mean travel distance exceed some critical lengths, could Equation 4.45 be applied to describe $\sigma_{\bar{x}}^2$.

With Equation 4.45, we can obtain the time-dependent dispersivity as follows:

$$\alpha_m = \frac{\alpha_0 \varepsilon_c^{1-D} D \bar{x}^{D-1} \left[1 - 2(D-1)\alpha_0 \varepsilon_c^{1-D} \bar{x}^{D-2} \right]}{\left[1 - D(D-1)\alpha_0 \varepsilon_c^{1-D} \bar{x}^{D-2} \right]^2} \tag{4.50}$$

Apparently, ε_c and $\bar{x}$ in Equation 4.50 must satisfy Equations 4.47 and 4.48, respectively, in order to ensure that the dispersivity is positive. It is easy to prove that α_m is a constant if D reduces to 1. On the other hand, if $D = 2$, Equation 4.45 shows that α_m grows linearly with $\bar{x}$, which is consistent with the Mercado model. Because Equation 4.45 is very complex, it is not convenient to use, and we may find a simple approximation. Assuming D is very close to 1, the second terms in the brackets of both the nominator and denominator may be omitted, and the expression reduces to:

$$\alpha_m = \alpha_0 \varepsilon_c^{1-D} D \bar{x}^{D-1} \tag{4.51}$$

If, on the contrary, D is very close to 2, the dispersivity can be estimated by:

$$\alpha_m = \frac{\alpha_0 \varepsilon_c^{1-D} D \bar{x}^{D-1}}{1 - D(D-1)\alpha_0 \varepsilon_c^{1-D} \bar{x}^{D-2}} \tag{4.52}$$

Generally speaking, if D is not very high, for example, $D < 1.3$, Equation 4.51 can give a good approximation.

Formally, Equation 4.45 is similar in appearance to the Wheatcraft-Tyler "single tube model." However, the two models are actually different. The difference lies in that the mean fractal travel length $\bar{x}_{fr}$ of all possible path lengths is used in the model presented here while the fractal length of the stream tube that covers a straight-line length of mean travel distance $\bar{x}$ was used in the Wheatcraft-Tyler model. As discussed above, the Wheatcraft-Tyler model tends to underestimate $\sigma_{\bar{x}}^2$. Furthermore, the model tends to significantly overestimate $\sigma_{\bar{x}}^2$.

Although our discussion here is focused on scale-dependency of dispersivity α, we are dealing, in fact, with time-dependency of α. The reason is that $\bar{x}$ is not an independent variable; rather, it is a function of the travel time t. In other words, $\bar{x}$ is simply a different expression of time and not with the x-axis coordinate. Further discussion on the difference between time-dependency and scale-dependency is beyond the scope of this chapter. Based on the above discussion, Equation 4.29 actually describes the relationship of $\sigma_{\bar{x}}^2$ versus time t. It is easy to show that the variance of travel distance may grow proportionally with the fractional power D of time with $D > 1$ instead of growing linearly with time t as commonly observed (Weeks, 1995). Therefore, the diffusion process described by Equation 4.29 is actually anomalous. To be specific, if σ_x^2 is proportional to t^D with $D > 1$, the diffusion process is referred to as superdiffusion.

4.5 Simulations

To show the effects of the fractal dimension on solute spreading, we conducted several simulations. First, we chose a value for the true constant dispersivity α_0. Here $\alpha_0 = 0.1$ m was used. Then, we determine the fractal cutoff limit ε_c according to Equation 4.47. In our example, ε_c is set to be 0.3 m. Now, we can calculate the variance of travel distance at different time (corresponding to the mean travel distance $\bar{x}$) given the fractal dimension of the stream lines is known. It should be pointed out that $\bar{x}$ should meet the requirements of Equation 4.48 and the calculated shortest travel distance must exceed the lower fractal cutoff limit. With $\bar{x}$ and σ_x^2 known, the solute distribution profile can be simulated if we assume the travel distance follows a normal distribution. Three examples are shown in Figures 4.7, 4.8, and 4.9. The figures show that enhanced diffusion occurs with the increase of the fractal dimension. If

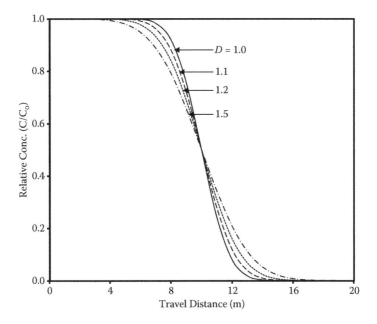

FIGURE 4.7
Comparison of solute profiles for systems with different fractal dimension at mean travel distance of 10 meters.

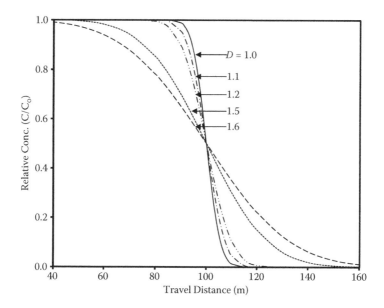

FIGURE 4.8
Comparison of solute profiles for systems with different fractal dimension at mean travel distance of 100 meters.

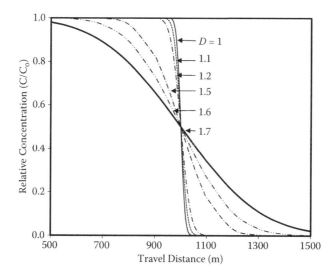

FIGURE 4.9
Comparison of solute profiles for systems with different fractal dimension at mean travel distance of 1000 meters.

the fractal dimension is close to 1, say, $D = 1.1$, the solute concentration profile does not show significant deviation from that of the nonfractal case ($D = 1$). On the contrary, the Wheatcraft-Tyler model predicts that the variance of travel distance increases proportionally to the mean travel distance raised to the power of $2D$. That is, for a system with D close to 1, the Wheatcraft-Tyler model would predict a spreading similar to that of the model presented here with D equal to 2. Hence, the Wheatcraft-Tyler model tends to overestimate the diffusion. In Figure 11 of Wheatcraft and Tyler (1988), they gave an example for the case of stream tubes having a fractal dimension of 1.0865. Their figure demonstrates that the diffusion is exceedingly exaggerated.

4.6 Application to Field-Scale Experiments

As shown in Equation 4.45, the relationship of variance of travel distance to mean travel distance is complicated. On the other hand, the two parameters true dispersivity α_0 and fractal cut-off limit ε_c are highly correlated. This makes it difficult for us to apply Equation 4.45 to field-scale data. Therefore, simplification of this formula is necessary. We consider the asymptotic behavior of Equation 4.45, that is, the case where the mean travel distance approaches infinity. It is easy to show that the limit of the denominator in Equation 4.45 is 1 given the mean travel distance approaches infinity if $D <$ 2. Therefore, we can say that σ_x^2 will grow proportionally to $\bar{x}$ raised to the power of D if the mean travel distance is large enough. It is easy to show that the reduced asymptotic form is actually the same as the single tube model in Wheatcraft and Tyler (1988) and is given by:

$$\sigma_x^2 = A\bar{x}^D \tag{4.53}$$

where $A = 2\alpha_0\varepsilon_c^{1-D}$ is a constant with dimension L^{2-D}. Equation 4.53 can be used to approximate the evolution of variance for field experiments. It should be pointed out that the power or exponent obtained is only close to the true fractal dimension D and not exactly the same because the asymptotic form is used to match the pre-asymptotic pattern. We use Equation 4.53 to describe several field-scale experiments. We chose the Cape Cod experiment (Garabedian et al., 1991), the Borden experiment (Rajaram and Gelhar, 1991), and the Columbus Air Force experiment (Adams and Gelhar, 1992). The fitted parameters for all three experiments are listed in Table 4.3. The value of the exponent D in all three cases is significantly different from 1.0, which suggests that all three sites express fractal behavior to some degree. The value of D for the Cape Cod site is smallest while that for the Columbus site is the largest and is very close to the upper limit of fractal dimension 2. If we assume the fractal dimension D is 1 for homogeneous media, then we conclude that the aquifer formation at the

TABLE 4.3

Fitted Parameters for Three Field-Scale Experiments

Site	Coefficient A	Exponent D
Cape Cod	1.1046 (0.2232)*	1.0989 (0.0251)
Borden	0.3447 (0.0783)	1.2068 (0.0315)
Columbus	1.8891 (0.2589)	1.8999 (0.0133)

*The value in parenthesis is the standard error of the estimation.

Columbus site is relatively more heterogeneous than those at the Cape Cod site and the Borden site. In effect, the aquifer at the Borden site is composed of clean, well-sorted, fine- to medium-grained sand and is quite homogeneous relative to many aquifers of similar origin (Mackay et al., 1986). The study area in the Cape Cod site is on a broad glacial outwash plain, and the upper 30 m of the aquifer is a medium to coarse sand with some gravel (Garabedian et al., 1991). The aquifer at the Columbus site is composed of poorly sorted to well-sorted sandy gravel and gravelly sand with minor amounts of silt and clay (Boggs et al., 1992). Furthermore, the variability in hydraulic conductivity at the Columbus site is greater than at other sites, including the Cape Cod and the Borden site (Adams and Gelhar, 1992). Therefore, the magnitude of the different D values given in Table 4.3 is in agreement with the observed heterogeneity of the three aquifers. The comparison of measured and fitted variance patterns is given in Figures. 4.10, 4.11, and 4.12. Generally, Equation

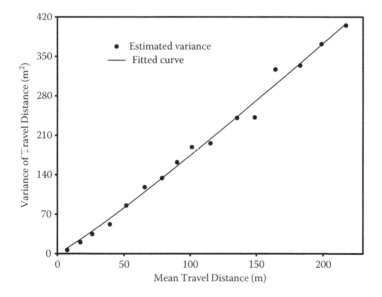

FIGURE 4.10
Comparison between estimated and fitted variance patterns at the Borden site.

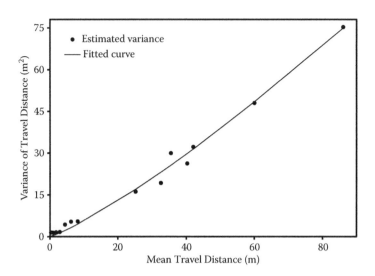

FIGURE 4.11
Comparison between estimated and fitted variance patterns at the Columbus site.

4.37 provides a good description of the variance pattern for all three sites, especially at large mean travel distance. For the Borden site experiment, the fitted curve underestimates the variance at the initial stage of transport. Perhaps, Equation 4.37 is not applicable to the initial stages of transport.

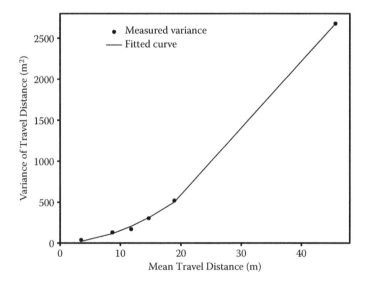

FIGURE 4.12
Comparison between estimated and fitted variance patterns at the cape Cod site.

4.7 Appendix A: Derivation of Finite-Difference Equations for CDE with a Linearly Time-Dependent Dispersivity

The governing equation for a linear dispersivity model reads (Equation 4.8):

$$\frac{\partial c}{\partial t} = -v\frac{\partial c}{\partial x} + a_1 v^2 t \frac{\partial^2 c}{\partial x^2} \tag{A1}$$

Denoting time and space increments by Δt and Δx, we can establish the finite difference scheme for point $(i\Delta x, j\Delta t)$, where i and j are integers and denote space and time steps, respectively. Finite difference approximations of each partial derivative in Equation A1 are as follows:

$$\frac{\partial c}{\partial t} = \frac{c_i^{j+1} - c_i^j}{\Delta t} \tag{A2}$$

$$\frac{\partial c}{\partial x} = \frac{c_{i+1}^{j+1} - c_i^{j+1}}{\Delta x} \tag{A3}$$

$$t\frac{\partial^2 c}{\partial x^2} = \frac{1}{2}\left[(j+1)\Delta t \frac{c_{i+1}^{j+1} - 2c_i^{j+1} + c_{i-1}^{j+1}}{\Delta x^2} + j\Delta t \frac{c_{i+1}^j - 2c_i^j + c_{i-1}^j}{\Delta x^2}\right] \tag{A4}$$

where c_i^j stands for solute concentration at node $(i\Delta x, j\Delta t)$. For convenience, let

$$\beta = \frac{\Delta t}{\Delta x}$$

Substituting Equations A2 through A4 into Equation A1 and rearranging gives:

$$Ac_{i-1}^{j+1} + Bc_i^{j+1} + Cc_{i+1}^{j+1} = E \tag{A5}$$

where

$$A = \frac{1}{2}a_1 v^2 (j+1)\beta^2 \tag{A6}$$

$$B = -1 + v\beta - a_1 v^2 (j+1)\beta^2 \tag{A7}$$

$$C = \frac{1}{2}a_1 v^2 (j+1)\beta^2 - v\beta \tag{A8}$$

$$E = -c_i^j - \frac{1}{2}a_1 v^2 j\beta^2 \left(c_{i+1}^j - 2c_i^j + c_{i-1}^j\right) \tag{A9}$$

After discretization, the upper boundary conditions (Equations 4.10 and 4.11) read:

$$[1 + a_1 v(j+1)\beta]c_1^{j+1} - a_1 v(j+1)\beta c_2^{j+1} = c_0, \quad 0 \le (j+1)\Delta t \le T \tag{A10}$$

$$[1 + a_1 v(j+1)\beta]c_1^{j+1} - a_1 v(j+1)\beta c_2^{j+1} = 0, \quad (j+1)\Delta t > T \tag{A11}$$

where c_0 is the solute concentration in input pulse, and T is the pulse duration.

4.8 Appendix B: Derivations of Finite-Difference Equations for CDE with a Linearly Distance-Dependent Dispersivity

In this case, the governing equation becomes (Equation 4.16):

$$\frac{\partial c}{\partial t} = a_2 v x \frac{\partial^2 c}{\partial x^2} - (1 - a_2)v \frac{\partial c}{\partial x} \tag{B1}$$

The finite-difference approximation of the above equation for point $(i\Delta x, j\Delta t)$ can be developed using the same notation as above. The approximation of the first partial derivatives of c with respect to time t and distance x are the same as Equations A2 and A3, respectively. The second derivative of concentration with respect to x is given by:

$$\frac{\partial^2 c}{\partial x^2} = \frac{1}{2}\left(\frac{c_{i+1}^{j+1} - 2c_i^{j+1} + c_{i-1}^{j+1}}{\Delta x^2} + \frac{c_{i+1}^{j} - 2c_i^{j} + c_{i-1}^{j}}{\Delta x^2} \right) \tag{B2}$$

Substituting Equation B2 together with Equations A2 and A3 into Equation B1, replacing x with $(i\Delta x)$, and rearranging gives an equation similar to Equation A5:

$$A'c_{i-1}^{j+1} + B'c_i^{j+1} + C'c_{i+1}^{j+1} = E' \tag{B3}$$

where

$$A' = \frac{1}{2}a_2 vi\beta \tag{B4}$$

$$B' = -1 + v\beta(1 - a_2) - a_2 vi\beta \tag{B5}$$

$$C' = \frac{1}{2} a_2 v i \beta - v \beta (1 - a_2) \tag{B6}$$

$$E' = -c_i^j - \frac{1}{2} a_2 v i \beta \left(c_{i+1}^j - 2 c_i^j + c_{i-1}^j \right) \tag{B7}$$

where β is defined as the ratio of time increment to space increment as above.

4.9 Appendix C: Derivations of Finite Difference Equations for CDE with a Nonlinearly Time-Dependent Dispersivity

For a power law dispersivity-time model, the governing equation is as follows (Equation 4.23):

$$\frac{\partial c}{\partial t} = -v \frac{\partial c}{\partial x} + a_3 v^{D_{fr}} t^{D_{fr}-1} \frac{\partial^2 c}{\partial x^2} \tag{C1}$$

Approximations of first partial derivatives are given in Equations A2 and A3. The second derivative with respect to x is given by:

$$t^{D_{fr}-1} \frac{\partial^2 c}{\partial x^2} = \frac{1}{2} \left[(j+1)^{D_{fr}-1} \Delta t^{D_{fr}-1} \frac{c_{i+1}^{j+1} - 2 c_i^{j+1} + c_{i-1}^{j+1}}{\Delta x^2} + (j \Delta t)^{D_{fr}-1} \frac{c_{i+1}^j - 2 c_i^j + c_{i-1}^j}{\Delta x^2} \right] \tag{C2}$$

For convenience, we let

$$\beta = \frac{\Delta t}{\Delta x} \quad \text{and} \quad \gamma = \frac{\Delta t^{D_{fr}}}{\Delta x^2}$$

Notice that $\gamma = \beta^2$ for $D_{fr} = 2$. Substituting Equation C2 together with Equations A2 and A3 into Equation C1 and rearranging yields:

$$A'' c_{i-1}^{j+1} + B'' c_i^{j+1} + C'' c_{i+1}^{j+1} = E'' \tag{C3}$$

where

$$A'' = \frac{1}{2} a_3 v^{D_{fr}} (j+1)^{D_{fr}-1} \gamma \tag{C4}$$

$$B'' = -1 + v\beta - a_3 v^{D_{fr}} (j+1)^{D_{fr}-1} \gamma \tag{C5}$$

$$C'' = \frac{1}{2} a_3 v^{D_{fr}} (j+1)^{D_{fr}-1} \gamma - v\beta \tag{C6}$$

$$E'' = -c_i^j - \frac{1}{2} a_3 v^{D_{fr}} j^{D_{fr}-1} \gamma \left(c_{i+1}^j - 2c_i^j + c_{i-1}^j \right) \tag{C7}$$

Obviously, Equation C3 reduces to Equation A5 for $D_{fr} = 2$.

For this case, the upper boundary conditions after discretization are given by:

$$\left[1 + a_3 v^{D_{fr}-1} (j+1)^{D_{fr}-1} \frac{\Delta t^{D_{fr}-1}}{\Delta x} \right] c_1^{j+1} - a_3 v^{D_{fr}-1} (j+1)^{D_{fr}-1} \frac{\Delta t^{D_{fr}-1}}{\Delta x} c_2^{j+1} = c_0,$$

$$0 \le (j+1)\Delta t \le T \tag{C8}$$

$$\left[1 + a_3 v^{D_{fr}-1} (j+1)^{D_{fr}-1} \frac{\Delta t^{D_{fr}-1}}{\Delta x} \right] c_1^{j+1} - a_3 v^{D_{fr}-1} (j+1)^{D_{fr}-1} \frac{\Delta t^{D_{fr}-1}}{\Delta x} c_2^{j+1} = 0,$$

$$(j+1)\Delta t > T \tag{C9}$$

where c_0 is the solute concentration in input pulse, and T is the pulse duration.

4.10 Appendix D: Derivations of Finite-Difference Equations for CDE with a Nonlinearly Distance-Dependent Dispersivity

For a parabolic dispersivity-distance model, the governing equation is (Equation 4.28):

$$\frac{\partial c}{\partial t} = a_4 v x^{D_{fr}-1} \frac{\partial^2 c}{\partial x^2} - \left[1 - a_4 (D_{fr} - 1) x^{D_{fr}-1} \right] v \frac{\partial c}{\partial x} \tag{D1}$$

For convenience, we let

$$\beta = \frac{\Delta t}{\Delta x}, \quad \xi = \frac{\Delta t}{\Delta x^{3-D_{fr}}}$$

Notice that $\xi = \beta$ for $D_{fr} = 2$. Replacing x with $(i\Delta x)$ and substituting approximation for partial derivatives, (e.g., Equations A2, A3, and B2), into the above equation and rearranging produces:

$$A'''c_{i-1}^{j+1} + B'''c_i^{j+1} + C'''c_{i+1}^{j+1} = E''' \tag{D2}$$

where

$$A''' = \frac{1}{2}a_4 v i^{D_{fr}-1} \xi \tag{D3}$$

$$B''' = -1 + v\beta - a_4 v\xi[i^{D_{fr}-1} + (D_{fr}-1)i^{D_{fr}-2}] \tag{D4}$$

$$C''' = \frac{1}{2}a_4 v\xi i^{D_{fr}-1} + a_4 v\xi(D_{fr}-1)i^{D_{fr}-2} - v\beta \tag{D5}$$

$$E''' = -c_i^j - \frac{1}{2}a_4 v\xi i^{D_{fr}-1}\left(c_{i+1}^j - 2c_i^j + c_{i-1}^j\right) \tag{D6}$$

Apparently, Equation D2 reduces to Equation B3 for $D_{fr} = 2$.

References

Adams, E. E., and L. W. Gelhar. 1992. Field study of dispersion in a heterogeneous aquifer. II. Spatial moments analysis. *Water Resour. Res.* 28(12): 3293–3307.

Arya, A., T. A. Hewett, R. Larson, and L. W. Lake. 1988. Dispersion and reservoir heterogeneity. Paper SPE14364 presented at the 60th Annual Technical Conference, Soc. of Pet. Eng., Las Vegas, Nev., Sept. 22–25, 1985.

Benson, D. A., S. W. Wheatcraft, and M. M. Meerschaert. 2000a. The fractional order governing equation of levy motion. *Water Resour. Res.* 36(6): 1413–1423.

Benson, D. A., S. W. Wheatcraft, and M. M. Meerschaert. 2000b. Application of a fractional advection-dispersion equation. *Water Resour. Res.* 36(6): 1403–1412.

Burns, E. 1996. Results of 2-dimensional sandbox experiments: Longitudinal dispersivity determination and seawater intrusion of coastal aquifers. Master's thesis, University of Nevada, Reno.

Boggs, J. M., S. C. Young, L. M. Beard, L. W. Gelhar, K. R. Rehfeldt, and E. E. Adams. 1992. Field study of dispersion in a heterogeneous aquifer. I. Overview and site description. *Water Resour. Res.* 28(12): 3281–3291.

Butters, G. L., and W. A. Jury. 1989. Field scale transport of bromide in an unsaturated soil. II. Dispersion modeling. *Water Resour. Res.* 25: 1583–1589.

Cushman, J. H., and T. R. Ginn. 1993. Nonlocal dispersion in media with continuously evolving scales of heterogeneity. *Transport in Porous Media* 13(1): 123–138.

Cushman, J. H., and T. R. Ginn. 2000. Fractional advection-dispersion equation: A classical mass balance with convolution-Fickian flux. *Water Resour. Res.* 36(12): 3763–3766.

Freyberg, D. L. 1986. A natural gradient experiment on solute transport in a sand aquifer. II. Spatial moments and the convection and dispersion of nonreactive tracers. *Water Resour. Res.* 22: 2031–2046.

Fried, J. J. 1972. Miscible pollution of ground water: A study of methodology. In *Proceedings of the International Symposium on Modelling Techniques in Water Resources Systems*, vol. 2, edited by A. K. Biswas, 362–371. Ottawa: Environment Canada.

Fried, J. J. 1975. *Groundwater Pollution: Theory, Methodology, Modeling and Practical Rules*. Amsterdam: Elsevier Scientific.

Garabedian, S. P., D. R. LeBlanc, L. W. Gelhar, and M. A. Celia. 1991. Large-scale natural gradient tracer test in sand and gravel, Cape Cod, Massachusetts. II. Analysis of spatial moments for a nonreactive tracer. *Water Resour. Res.* 27(5): 911–924.

Gelhar, L. W., C. Welty, and K. R. Rehfeldt. 1992. A critical review of data on field-scale dispersion in aquifers, *Water Resour. Res.* 28(7): 1955–1974.

Green, P. E., and J. D. Caroll. 1978. *Analyzing Multivariate Data*. New York: John Wiley & Sons.

Jury, W. A., and K. Roth. 1990. *Transfer Functions and Solute Movement through Soil: Theory and Applications*. Basel: Birkhauser Verlag.

Khan, A. U.-H., and W. A. Jury. 1990. A laboratory study of the dispersion scale effect in column outflow experiments. *J. Contaminant Hydrol. Amsterdam* 5: 119–131.

Liu, H. H., and J. H. Dane. 1996. An extended transfer function model of field-scale solute transport: Model development. *Soil Sci. Soc. Am. J.* 69: 986–991.

Logan, J. D. 1996. Solute transport in porous media with scale-dependent dispersion and periodic boundary conditions. *J. Hydrol.* 184: 261–276.

Mackay, D. M., D. L. Freyberg, P. V. Roberts, and J. A. Cherry. 1986. A Natural Gradient Experiment on Solute Transport in a Sand Aquifer 1. Approach and Overview of Plume Movement. *Water Resour. Res.*, 22(13): 2017–2029.

Mercado, A. 1967. *The Spreading Pattern of Injected Water in a Permeability-Stratified Aquifer*. IAHS AISH Publ., Vol. 72, 23–36.

Neuman, S. P. 1990. Universal scaling of hydraulic conductivities and dispersivities in geologic media. *Water Resour. Res.* 26(8): 1749–1758.

Neuman, S. P., and Y.-K. Zhang. 1990. A quasi-linear theory of non-Fickian and Fickian subsurface dispersion. I. Theoretical analysis with application to isotropic media. *Water Resour. Res.* 26(5): 887–902.

Pachepsky, Y., D. Benson, and W. Rawls. 2000. Simulating scale-dependent solute transport in soils with the fractional advective-dispersive equation. *Soil Sci. Soc. Am. J.* 64: 1234–1243.

Pang, L., and M. Close. 1999. Field-scale physical non-equilibrium transport in an alluvial gravel aquifer. *J. Contaminant Hydrol.* 38: 447–464.

Peaudecerf, P., and J. P. Sauty. 1978. Application of a mathematical model to the characterization of dispersion effects of groundwater quality. *Prog. Water Technol.*, 10(5/6): 443–454.

Pickens, J. F., and G. E. Grisak. 1981a. Scale-dependent dispersion in a stratified granular aquifer. *Water Resour. Res.* 17(4): 1191–1211.

Pickens, J. F., and G. E. Grisak. 1981b. Modeling of scale-dependent dispersion in hydrogeologic systems. *Water Resour. Res.* 17(6): 1701–1711.

Podlubny, I. 1999. *Fractional Differential Equations*. San Diego, CA: Academic Press.

Porro, I., P. J. Wierenga, and R. G. Hills. 1993. Solute transport through large uniform and layered soil columns. *Water Resour. Res.* 29: 1321–1330.

Press, W. H., S. A. Teukolsky, W. T. Vetterling, and B. P. Flannery. 1992. *Numerical Recipes in FORTRAN: The Art of Scientific Computing*, second edition. New York: Cambridge University Press.

Rajaram, H., and L. W. Gelhar. 1991. Three-dimensional spatial moments analysis of the Borden tracer test. *Water Resource Research*, Vol. 27(6): 1239–1251.

SAS Institute Inc. 2001. The SAS System for Windows, Release 8.2. SAS Institute Inc. Cary, NC.

Su, N.-H. 1995. Development of the Fokker-Planck equation and its solutions for modeling transport of conservative and reactive solutes in physically heterogeneous media. *Water Resour. Res.* 31(12): 3025–3032.

Sudicky, E. A., J. A. Cherry, and E. O. Frind. 1983. Migration of contaminants in groundwater at a landfill: A case study. IV. A natural-gradient dispersion test. *J. Hydrol.* 63: 81–108.

Sudicky, E. A., and J. A. Cherry. 1979. Field observations of tracer dispersion under natural flow conditions in an unconfined sandy aquifer. *Water Pollut. Res. Can.* 14: 1–17.

Taylor, S. R., and K. W. F. Howard. 1987. A field study of scale-dependent dispersion in a sandy aquifer. *J. Hydrol.* 90: 11–17.

Weeks, E. R., T. H. Solomon, J. S. Urbach, and H. L. Swinney, 1995. Observation of anomalous diffusion and Levy flights. In Michael F. Shlesinger, George M. Zaslavsky, and Uriel Frisch (Eds.) Levy Flights and Related Topics in Physics, Springer, 1995, Pp.51-71.

Wheatcraft, S. W., and S. W. Tyler. 1988. An explanation of scale-dependent dispersivity in heterogeneous aquifers using concepts of fractal geometry. *Water Resour. Res.* 24(4): 566–578.

Xu, M., and Y. Eckstein. 1995. Use of weighted least-squares method in evaluation of the relationship between dispersivity and field scale. *Ground Water* 33(6): 905–908.

Yates, S. R. 1990. An analytical solution for one-dimension transport in heterogeneous porous media. *Water Resour. Res.* 26: 2331–2338.

Yates, S. R. 1992. An analytical solution for one-dimension transport in porous media with an exponential dispersion function. *Water Resour. Res.* 28: 2149–2154.

Zhang, R. 2000. Generalized transfer function model for solute transport in heterogeneous soils. *Soil Sci. Soc. Am. J.* 64: 1595–1602.

Zhang, R., K. Huang, and J. Xiang. 1994. Solute movement through homogeneous and heterogeneous soil columns. *Adv. Water Resour.* 17: 317–324.

Zhang, Y.-K., and S. P. Neuman. 1990. A quasi-linear theory of non-Fickian and Fickian subsurface dispersion. II. Application to anisotropic media and the Borden site. *Water Resour. Res.* 26(5): 903–913.

Zhou, L.-Z., and H. M. Selim. 2002. A conceptual fractal model for describing time-dependent dispersivity. *Soil Sci.* 167(3): 173–183.

Zhou, L. and H M Selim. 2003. Scale-Dependent Dispersion in Soils: An Overview. *Adv. Agron.* 80: 321–263.

Zou, S., J. Xia, and A. D. Koussis. 1996. Analytical solutions to non-Fickian subsurface dispersion in uniform groundwater flow. J. Hydrol. 179: 237–258.

5

Multiple-Reaction Approaches

Several studies showed that sorption-desorption of dissolved chemicals on different soils was not adequately described by use of a single reaction of the equilibrium or kinetic type. Failure of single reactions is not surprising since they only describe the behavior of one species, with no consideration given to the simultaneous reactions of others in the soil system. Multisite or multireaction models deal with the multiple interactions of any one species in the soil environment. Such models are empirical in nature and are based on the assumption that retention sites are not homogeneous in nature; rather, the sites are heterogeneous and thus have different affinities to individual solute species. Figure 5.1 clearly illustrates the heterogeneous make up of the soil matrix.

It is well recognized that multireaction models cannot account for all possible interactions occurring in the soil-water environment. For example, characterization of chemical, biological, and physical interactions of nitrogen within the soil environment is a prerequisite in the formulation of a multireaction nitrogen model. One major point is how strongly such factors affect nitrogen behavior and distribution within soil systems. Among these factors are the effect of soil texture and structure on oxygen diffusion; distribution of plant residues (vertically and horizontally), which affects infiltration rates, leaching, and biological transformations, including plant uptake, mineralization, and denitrification; and cultural practices such as tillage and fertilizer distribution, which affect nitrogen distribution vertically and horizontally. Since nitrate is a highly mobile nitrogen form that can leach through the soil profile and eventually into groundwater, the goal of any nitrogen management plan must include minimizing nitrate leaching from agricultural activities into groundwater. Nevertheless, description of N dynamics in the soil is often simplified. To illustrate this, one needs to examine the N dynamics in soils as described in the simplified schematic for N transformations shown Figure 5.2. The model accounts for nitrification, denitrification, immobilization, mineralization, and ion exchange of ammonium as a reversible first-order kinetic process. It also accounts for uptake by plant roots and transport in the soil water.

Based on first-order kinetic reactions governing N transformations, a set of simultaneous equations describing the N cycle were derived by Mehran and Tanji (1974). This set of equations was solved simultaneously to provide N simulations. Wagenet, Biggar, and Nielseen (1977) found excellent agreement between observed urea, ammonium, and nitrate effluent concentrations and

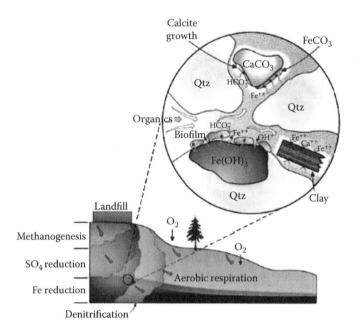

FIGURE 5.1

Schematic representation of the oxidation-reduction zones that may develop in an aquifer downstream from an organic-rich landfill. Closest to a landfill may be a zone of methanogenesis, which is progressively followed downstream by zones of sulfate reduction, dissimilatory iron reduction, denitrification, and aerobic respiration that develop as the plume becomes progressively oxidized through the influx of oxygenated water. Within the dissimilatory iron reduction zone, a pore scale image is shown in which the influx of dissolved organics provides electrons for dissimilatory iron reduction mediated by a biofilm. The dissolution of this phase leads to the release of Fe^{2+}, HCO_3^-, and OH^- into the pore fluid, potentially driving siderite or calcite precipitation downstream, and thus reducing the porosity and permeability of the material. Sorption of Fe^{2+} may also occur on clays, displacing other cations originally present on the mineral surface. Where reactions are fast relative to local transport, gradients in concentration, and thus in reaction rates, may develop at the pore scale. (From C. I. Steefel et al. 2005. *Earth and Planetary Science Letters* 240: 539–558. With permission.)

theoretical results. Urea hydrolysis, nitrification, and dentrification were assumed to be governed by first-order kinetics as described above.

Unlike the N cycle, where transformation processes have been identified for several decades, for other reactive chemicals, the governing processes of sorption are at best ill defined. For example, reactive chemical transport in soils may be complicated by precipitation reactions that act as kinetic sinks for chemicals. Because of its agronomic and environmental significance, P precipitation reactions during transport in soils have been most frequently studied. Often P is a severely limited nutrient in terrestrial and aquatic ecosystems, which adversely affects plant and fish production. Because of the significance of P precipitation reactions in soil, the geochemical

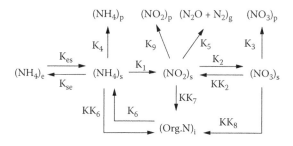

FIGURE 5.2
A schematic representation of transformation of soil nitrogen. Terms K and KK refer to rate coefficients and e, s, p, I, and g refer to exchangeable, solution, immobilized, and gaseous phases, respectively. (From M. Mehran and K. K. Tanji. 1974. *J. Environ. Qual.* 3: 391–395. With permission.)

and hydrologic processes controlling P transport must be clearly understood. Many of the principles to be discussed for P reactions may in some instances be applicable to other types of precipitation reactions (i.e., Al, Fe, etc.). Mansell et al. (1977), investigating the transport of orthophosphate through saturated and unsaturated columns of a sandy soil, found that a single process failed to describe P transport. Specifically, reversible equilibrium adsorption-desorption relationship of the Freundlich type inadequately described observed data. By coupling a first-order kinetic expression with the classical transport equation and considering nonlinear exchange of the Freundlich type, Mansell et al. (1977) substantially improved the prediction of orthophosphate transport through soil (Figure 5.3). They noted, however, that this model overpredicted the peak

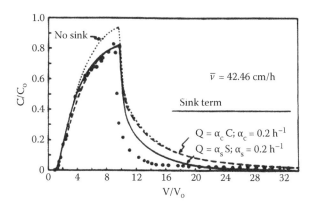

FIGURE 5.3
Observed phosphate effluent condensations from the Al horizon of a sandy soil with predicted curves determined using a one-site, nonlinear, nonequilibrium model with and without a sink term for irreversible sorption and immobilization.

concentration and tailing response of observed P effluent concentrations. They further modified their model by considering first-order irreversible precipitation described by:

$$Q = \alpha_c c$$

and first-order irreversible chemical immobilization via physical adsorption described by:

$$Q = \alpha_s S$$

where α_c and α_s (h^{-1}) are the rate coefficients for precipitation and chemical immobilization, respectively, and Q is the rate of solute consumption (sink). Incorporating an irreversible sink for chemical immobilization or precipitation into the convective-dispersive transport equation (CDE) provided significantly better agreement between the observed data and the model-predicted curve (Figure 5.3).

5.1 Two-Site Models

One of the earliest multireaction models is the two-site model proposed by Selim, Davidson, and Mansell (1976). This model was developed to describe observed batch results, which showed rapid initial retention reactions followed by slower retention reactions. The model was also developed to describe the excessive tailing of breakthrough curves (BTCs) obtained from pulse inputs in miscible displacement experiments. The two-site model is based on several simplifying assumptions. First, it is assumed that a fraction of the total sites (referred to as type I sites) reacts rapidly with the solute in soil solution. In contrast, we assume that type II sites are highly kinetic in nature and react slowly with the soil solution. The retention reactions for both types of sites were based on the nonlinear (or nth order) reversible kinetic approach and may be expressed as:

$$\frac{\partial S_1}{\partial t} = k_1 \frac{\Theta}{\rho} C^n - k_2 S_1 \tag{5.1}$$

$$\frac{\partial S_2}{\partial t} = k_3 \frac{\Theta}{\rho} C^m - k_4 S_2 \tag{5.2}$$

$$S_T = S_1 + S_2 \tag{5.3}$$

where S_1 and S_2 are the amounts of solute retained by sites I and sites II, respectively, S_T is the total amount of solute retained by the soil matrix (mg kg^{-1}), and k_1, k_2, k_3, and k_4 are the associated rate coefficients (h^{-1}). The nonlinear parameters n and m are considered to be less than unity and $n \neq m$. For the case $n = m = 1$, the retention reactions are of the first-order type and the problem becomes a linear one.

This two-site approach was also considered for the case where type I sites were assumed to be in equilibrium with the soil solution, whereas type II sites were considered to be of the kinetic type. Such conditions may be attained when the values of k_1 and k_2 are extremely large. Under these conditions, a combination of equilibrium and kinetic retention is (Selim, Davidson, and Mansell, 1976):

$$S_1 = K_f C^n \tag{5.4}$$

$$\frac{\partial S_2}{\partial t} = k_3 \frac{\Theta}{\rho} C^m - k_4 S_2 \tag{5.5}$$

Jardine, Parker, and Zelazny (1985) found that the use of the equilibrium and kinetic two-site model provided good predictions of BTCs for Al in kaolinite at different pH values. Selim, Davidson, and Mansell (1976) found that the two-site model yielded improved predictions of the excessive tailing of the desorption or leaching side and the sharp rise of the sorption side of the BTCs in comparison with predictions using single-reaction equilibrium or kinetic models. The two-site model has been used by several scientists, including DeCamargo, Biggar, and Nielsen (1979), Nkedi-Kizza et al. (1984), Jardine, Parker, and Zelazny (1985), and Parker and Jardine (1986), among others. The model proved successful in describing the retention and transport of several dissolved chemicals, including Al, P, K, Cr, Cd, 2,4-D, atrazine, and methyl bromide. However, there are some inherent disadvantages of the two-site model. First, the reaction mechanisms are restricted to those that are fully reversible. Moreover, the model does not account for possible consecutive-type solute interactions in the soil system.

5.2 Multireaction Models

A schematic representation of the multireaction model is shown in Figure 5.4. In this model we consider the solute to be present in the soil solution phase (C) and in five phases representing solute retained by the soil matrix as S_e, S_1, S_2, S_3, and S_{irr}. We further assume that S_e, S_1, and S_2 are in direct contact with

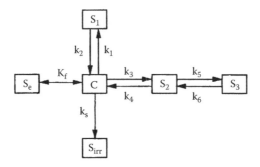

FIGURE 5.4
A schematic representation of the multireaction model.

the solution phase and are governed by concurrent type reactions. Here we assume S_e is the amount of solute that is sorbed reversibly and is in equilibrium with C at all times. Specifically, the multireaction model assumes that the total amount sorbed S_T or simply S is the total amount retained reversibly or reversibly by soil matrix surfaces, $S = S_e + S_1 + S_2 + S_3 + S_{irr}$.

The governing equilibrium retention/release mechanism is the nonlinear Freundlich type:

$$S_e = K_f C^b \tag{5.6}$$

where K_f is the associated distribution coefficient ($cm^3\ kg^{-1}$) and b is a dimensionless Freundlich parameter ($b < 1$). Other equilibrium-type retention mechanisms are given in Chapter 3. These include linear ($b = 1$), Langmuir, two-site Langmuir, and others.

The retention/release reactions associated with S_1, S_2, and S_3 are concurrent- or consecutive-type kinetic reactions. Specifically, the S_1 and S_2 phases were considered to be in direct contact with C and reversible rate coefficients of the (nonlinear) kinetic type govern their reactions:

$$\frac{\partial S_1}{\partial t} = k_1 \frac{\Theta}{\rho} C^n - k_2 S_1 \tag{5.7}$$

$$\frac{\partial S_2}{\partial t} = k_3 \frac{\Theta}{\rho} C^m - (k_4 + k_5) S_2 + k_6 S_3 \tag{5.8}$$

$$\frac{\partial S_3}{\partial t} = k_5 S_2 - k_6 S_3 \tag{5.9}$$

where k_1 and k_2 are the forward and backward rate coefficients (h^{-1}), respectively, and n is the reaction order associated with S_1. Similarly, k_3 and k_4 are the rate coefficients and m is the reaction order associated with S_2, and k_5 and k_6 are the reaction parameters associated with S_3. In the absence of the consecutive reaction between S_2 and S_3, that is, if $S_3 = 0$ at all times ($k_5 = k_6 = 0$), Equation 5.8 reduces to:

$$\frac{\partial S_2}{\partial t} = k_3 \frac{\Theta}{\rho} C^m - k_4 S_2$$

(5.10)

Thus, Equation 5.10 for S_2 resembles that for S_1 except for the magnitude of the associated parameters k_3, k_4, and m.

The sorbed phases (S_e, S_1, S_2, S_3) may be regarded as the amounts sorbed on surfaces of soil particles and chemically bound to Al and Fe oxide surfaces or other types of surfaces, although it is not necessary to have a priori knowledge of the exact retention mechanisms for these reactions to be applicable. These phases may be characterized by their kinetic sorption and release behavior to the soil solution and thus are susceptible to leaching in the soil. In addition, the primary difference between these two phases not only lies in the difference in their kinetic behavior but also in the degree of nonlinearity as indicated by the parameters n and m. The sink/source term Q is commonly used to account for irreversible reactions such as precipitation/dissolution, mineralization, and immobilization, among others. We express the sink term as a first-order kinetic process:

$$Q = \rho \frac{\partial S_3}{\partial t} = k_s \Theta C$$

(5.11)

where k_s is the associated rate coefficient (h^{-1}). The sink term Q is expressed in terms of a first-order irreversible reaction for reductive sorption or precipitation or internal diffusion as described by Amacher et al. (1986) and Amacher, Selim, and Iskandar (1988). Equation (5.11) is similar to that for the diffusion-controlled precipitation reaction if one assumes that the equilibrium concentration for precipitation is negligible.

This model is multipurpose in nature, which accounts for linear as well as nonlinear reaction processes of the reversible and irreversible types. The capability of the model is not limited to describing commonly measured batch-type sorption data (following a specific reaction time, e.g., 1 day) but also in describing changes in concentration with time of reaction during sorption as well as desorption. Therefore, the uniqueness of this model is that its aim is to describe the reactivity of solutes with natural systems versus time during sorption or desorption. In contrast, for most models, for example, the simple linear, Freundlich, Langmuir, DDRM (double-domain reaction model), and TDRM (treble-domain reaction model), two distinct sets of parameters are obtained, one for adsorption and one for desorption (Weber, McGinley, and Katz, 1992; Lesan and Bhandari, 2003). Moreover, the use of such models yields a set of parameters that are only applicable for a specific reaction time. On the other hand, the multireaction model presented here provides a comprehensive accounting of the sorption-desorption processes, where a single set of parameters is applicable for an entire data set and for a wide range of initial (or input) concentrations.

In order to describe transport of reactive chemicals in soils and geological media, it is necessary to incorporate the multireaction model equations described above with the CDE described in Chapter 3. For steady water flow conditions, the resulting CDE equation becomes

$$\frac{\partial C}{\partial t} + \frac{\rho}{\theta}\frac{\partial S}{\partial x} = D\frac{\partial^2 C}{\partial x^2} - v\frac{\partial C}{\partial x} \tag{5.12}$$

where x is distance (cm), D is dispersion coefficient (cm^2 h^{-1}), v ($=q/\theta$) is average pore water velocity (cm h^{-1}), and q is Darcy's water flux density (cm h^{-1}). The appropriate initial and boundary conditions for a finite soil column are

$$C(x) = C_{init} \quad t = 0 \tag{5.13}$$

$$S(x) = S_{init} \quad t = 0 \tag{5.14}$$

$$\left(-D\frac{\partial C}{\partial x} + vC\right)\Bigg|_{x=0} = \begin{cases} vC_o & t \in T_p \\ 0 & t \notin T_p \end{cases} \tag{5.15}$$

$$\frac{\partial C}{\partial x}\Bigg|_{x=L} = 0 \quad t > 0 \tag{5.16}$$

where C_{init} is the initial solution concentration (mg L^{-1}), S_{init} is the initial amount of sorption (mg kg^{-1}), C_o is the input solute concentration (mg L^{-1}), T_p is the duration of applied solute pulses, and L is the length of the column (cm). The above Equations (5.12 to 5.16) were solved numerically using the finite–difference, Crank-Nicholson explicit-implicit approximation, which was discussed in Chapter 3.

5.3 Applications

Multireaction models have been applied successfully to describe the transport and retention of numerous reactive chemicals in soils and geological media, including trace elements and heavy metals, radionuclides, military explosives, herbicides, and pesticides. Selected examples are presented in this section.

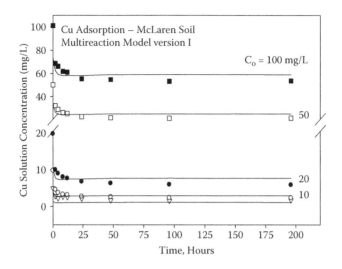

FIGURE 5.5
Experimental adsorption results of Cu from soil solution in McLaren soil versus time for several initial concentrations (C_o). The solid curves were obtained using the multireaction model with S_1 and S_{irr}. Estimated model parameters along with their standard errors (k_1, k_2, and k_3) are given in Table 5.1.

Figure 5.5 shows experimental adsorption results of Cu in soil solution in McLaren soil versus time for several initial concentrations (C_o). The solid curves were obtained using the multireaction model shown in Figure 5.4 (see Selim and Ma, 2001). The use of this model was limited to two sorbed phases, one reversible (S_1) and the other irreversible (S_{irr}). This was carried out by setting up all parameters associated with the remaining phases as zero. Estimated model parameters along with their standard errors (k_1, k_2, and k_{irr}) are given in Table 5.1. Here the estimated parameters were independently obtained for each initial concentration (C_o) as well as for the entire data set

TABLE 5.1

Goodness of Fit of the Multireaction Model for Individual and Overall Initial Concentrations (C_o)

C_o (mg/L)	r^2	RMSE	k_1 (h^{-1})	SE (h^{-1})	K_2 (h^{-1})	SE (h^{-1})	K_{irr} (h^{-1})	S_E (h^{-1})
5	0.601	0.3601	0.682	0.225	0.429	0.167	0.0062	0.0022
10	0.849	0.3668	0.922	0.109	0.270	0.043	0.0079	0.0022
20	0.762	0.8887	1.133	0.179	0.315	0.063	0.0059	0.0018
50	0.802	2.1709	0.994	0.163	0.273	0.056	0.0028	0.0010
100	0.663	4.3336	1.217	0.252	0.357	0.087	0.0019	0.0007
Overall	0.992	2.0635	1.131	0.089	0.323	0.031	0.0020	0.0003

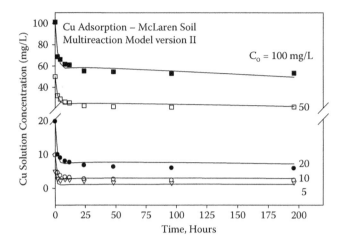

FIGURE 5.6
Experimental adsorption results of Cu from soil solution in McLaren soil versus time for several initial concentrations (C_o). The solid curves were obtained using the multireaction model with S_1 and S_s.

(for all values of C_o) using parameter optimization (see Table 5.1). Reduced standard errors for the model parameters k_1, k_2, and k_{irr} were realized when the entire data set is used. Of significance is the fact that one set of model parameters can thus be used for the entire concentration range, which lends credence to the model in describing the Cu.

To further test the applicability of the multireaction model, the solid curves shown in Figure 5.6 were obtained based on a different version of the model. Specifically, this model version accounted for two sorbed phases, one reversible (S_2) and the other irreversible (S_s). It is obvious from the simulations shown in Figures 5.5 and 5.6 that a number of model versions are capable of producing indistinguishable simulations of the data. Similar conclusions were arrived at by Amacher, Selim, and Iskandar (1988, 1990) for Cd, Cr(VI), and Hg for several soils. They also stated that it was not possible to determine whether the irreversible reaction is concurrent or consecutive, since both model versions provided similar fit of their batch data. This finding implies that the adsorption data for Cu alone is insufficient and that additional data are needed to arrive at a recommendation for the adoption of a specific model version.

In Figure 5.7, results of mercury (Hg) transport in a sand column are presented along with multireaction model simulations (see Liao, Delaune, and Selim, 2009). This is one of the few studies that indicate significant Hg mobility in soil columns. (See also the results of Miretzky, Bisinoti, and Jardim, 2005, on alluvial, podozol and humic gley soils from the Amazon region.) The Hg input pulse concentration used here was 8 mg/L $Hg(NO_3)_2$ prepared

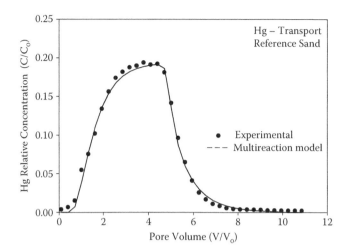

FIGURE 5.7

Experimental Hg (II) breakthrough curve (BTC) from the reference sand column (C_o = 8 mg L^{-1}). The solid curve is multireaction model simulation where the rates of reactions were 0.340 ± 0.019, 0.033 ± 0.001, and 0.001 ± 0.001 h^{-1} for k_1, k_2, and k_3, respectively.

in 0.01 M Ca(NO$_3$)$_2$ background solution. In the sand column, a noticeable Hg peak was observed and did not exceed 2 mg/L (see Figure 5.7). In fact, for the BTC of the reference sand, which exhibited symmetry, the recovery of Hg from the soil column was only 17.3% of that applied. Therefore, more than 80% of the applied Hg was strongly retained by the reference sand column. Recently, Wernert, Frimmel, and Behra (2003) reported strong Hg retention in a column experiment of quartz sand (99% quartz and amorphous silica) where continuous Hg pulse application of 100 mg/L was maintained. In fact, no Hg was observed in the column effluent during the first 100 pore volumes. Some 500 pore volumes of Hg application were needed before a concentration maximum of 22 mg/L was reached. Wernert, Frimmel, and Behra (2003) did not report the percent of Hg retained in their quartz column.

The mercury BTC for the reference sand column was successfully described using the multireaction model. Liao, Delaune, and Selim (2009) concluded that a simple model formulation with reversible kinetic and irreversible sites (S_1 and S_{irr}) is recommended for the case of this sand column. Here the irreversible reaction associated with S_i may be considered as inner-sphere complexation as suggested by Sarkar, Essington, and Misra (1999). Interparticle diffusion is another process that is responsible for retention of Hg. Such a process is often considered as a rate-limiting step (see Yin, Allen, and Huang, 1997; Miretzky, Bisinoti, and Jardim, 2005). Nevertheless, model validation and verification are needed, which require further experimental investigation of the processes associated with Hg sorption and transport in soils.

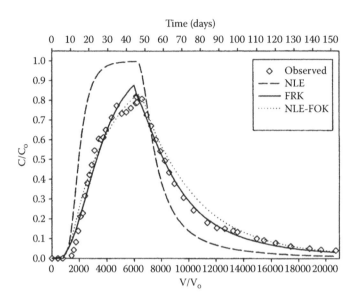

FIGURE 5.8

Observed and model calculated breakthrough curves for U(VI) in Hanford (HF) column. Pulse input $C_o = 0.29$ mg/L, pH = 6.8 in 0.01 M NaNO$_3$. The solid curve is based on the multireaction model with only S_1 considered; the dotted curve shows S_e and S_1 reversible phases. The dashed curve is based on the assumption of nonlinear (Freundlich) equilibrium.

Figure 5.8 shows breakthrough results from a column of Hanford soil where a pulse input of uranium (U(VI)) in a 0.01 M NaNO$_3$ background matrix was introduced to each of the columns at a flow rate of 4.3 cm/h (Barnett et al., 2000). After the pulse, the inlet solution was switched to a U(VI)-free, 0.01 M NaNO$_3$ solution. The observed results indicate that U(VI) is highly mobile where total recovery was attained. The solid curve is based on model calculations where only S_1 was considered, whereas the dotted curve is where S_e and S_1 reversible phases were used. The dashed curve is based on the assumption of full equilibrium with S_e only. Therefore, a combination of fully reversible equilibrium and kinetic reactions best describe the U(VI) transport in Hanford soil.

Transport results of two military explosives, TNT and RDX, are illustrated by the BTCs shown in Figure 5.9 (Selim, Xue, and Iskandar, 1995). A pulse having a concentration of either 10 or 100 mg/L TNT or RDX dissolved in 0.005 M Ca(NO$_3$)$_2$ solution was introduced into each column at the same flux. The applied pulse was subsequently followed by several pore volumes of the background solution. In comparison with the tracer tritium, RDX exhibited high mobility and little tailing. The solid curve is

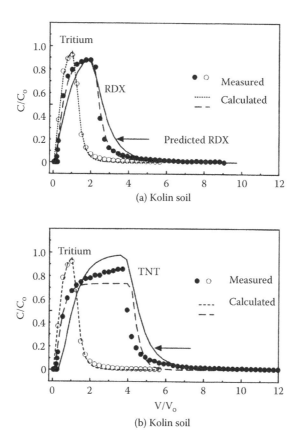

FIGURE 5.9

Transport of tritium, RDX, and TNT in Kolin soil. Solid curves are simulations based on the multireaction model.

based on multireaction model calculations where batch kinetic parameters were used. For TNT, the use of kinetic batch parameters overestimated the extent of retardation. Specifically batch parameters overestimated the extent of retardation, which is probably because the mixing was more complete in batch experiments than in the miscible displacement study. For several organic contaminants, several authors postulated that experimental artifacts of batch experiments were responsible for the discrepancy. The dashed curves are model simulations that provided good description of RDX and TNT transport results. In obtaining these simulations, only two phases were considered in the multireaction model, an equilibrium (S_e) phase and an irreversible phase (S_{irr}).

References

Amacher, M. C., J. Kotuby-Amacher, H. M. Selim, and I. K. Iskandar. l986. Retention and release of metals by soil-evaluation of several models. *Geoderma* 38: 131–154.

Amacher, M. C., H. M. Selim, and I. K. Iskandar. 1988. Kinetics of chromium (VI) and cadmium retention in soils: A nonlinear multireaction model. *Soil Sci. Soc. Am. J.* 52:398–408.

Amacher, M. C., H. M. Selim, and I. K. Iskandar. 1990. Kinetics of mercuric chloride retention in soils. *J. Environ. Qual.* 19: 382–388.

Barnett, M. O., P. M. Jardine, S. C. Brooks, and H. M. Selim. 2000. Adsorption and transport of uranium (VI) in subsurface media. *Soil Sci. Soc. Am. J.* 64: 908–917.

DeCamargo, O. A., J. W. Biggar, and D. R. Nielsen. 1979. Transport of inorganic phosphorus in an Alfisol. *Soil Sci. Soc. Am. J.* 43: 884–890.

Jardine, P. M., J. C. Parker, and L. W. Zelazny. 1985. Kinetics and mechanisms of aluminum adsorption on kaolinite using a two-site nonequilibrium transport model. *Soil Sci. Soc. Am. J.* 49: 867–873.

Lesan, H. M., and A. Bhandari. 2003. Atrazine sorption on surface soils: Time-dependent phase distribution and apparent desorption hysteresis. *Water Res.* 37: 1644–1654.

Liao, L., R. D. Delaune, and H. M. Selim. 2009. Mercury adsorption-desorption and transport in soils. *J. Environ. Qual.* 38: 1608–1616.

Nkedi-Kizza, P., J. M. Biggar, H. M. Selim, M. Th. van Genuchten, P. J. Wierenga, J. M. Davidson, and D. R. Nielsen. 1984. On the equivalence of two conceptual models for describing ion exchange during transport through an aggregated soil. *Water Resour. Res.* 20: 1123—1130.

Mansell, R. S., H. M. Selim, P. Kanchanasut, J. M. Davidson, and J. G. A. Fiskell. 1977. Experimental and simulated transport of phosphorus through sandy soils. *Water Resour. Res.* 13: 189–194.

Mehran, M., and K. K. Tanji. 1974. Computer modeling of nitrogen transformations in soils. *J. Environ. Qual.* 3: 391–395.

Miretzky, P., M. C. Bisinoti, and W. F. Jardim. 2005. Sorption of mercury (II) in Amazon soils from column studies. *Chemosphere* 60: 1583–1589.

Parker, J. C., and P. M. Jardine. 1986. Effect of heterogeneous adsorption behavior on ion transport. *Water Resour. Res.* 22: 1334–1340.

Sarkar, D., M. E. Essington, and K. C. Misra. 1999. Adsorption of mercury (II) by variable charge surface of quartz and gibbsite. *Soil Sci. Soc. Am. J.* 63: 1626–1636.

Selim, H. M., J. M. Davidson, and R. S. Mansell. 1976. Evaluation of a two site adsorption-desorption model for describing solute transport in soils. In Proceedings of the Summer Computer Simulation Conference, Washington, D. C., Simulation Councils, La Jolla, CA, 444–448.

Selim, H. M., and L. Ma. 2001. Modeling nonlinear kinetic behavior of copper adsorption-desorption in soil. *Soil Sci. Soc. Am. Spec. Publ.* 56: 189–212.

Selim, H. M., S. K. Xue, and I. K. Iskandar. 1995.Transport of 2,4,6-trinitrotoluene and hexahydro-1,3,5-trinitro-1,3,5-triazine in soils. *Soil Sci.* 160: 328–339.

Steefel, C. I., D. J. DePaolo, and P. C. Lichtner. 2005. Reactive transport modeling: An essential tool and a new research approach for the Earth sciences. *Earth Planetary Sci. Lett.* 240: 539–558.

Yin, Y., H. E. Allen, and C. P. Huang. 1997. Kinetics of mercury (II) adsorption and desorption on soil. *Environ. Sci. Technol.* 31: 496–503.

Wagenet, R. J., J. W. Biggar, and D. R. Nielseen. 1977. Tracing the transformation of urea fertilizer during leaching. *Soil Sci. Soc. Am. J.* 41: 473–480.

Weber, Jr.,W. J., P. M. McGinley, and L. E. Katz. 1992. A distributed reactivity model for sorption by soils and sediments. 1. Conceptual basis and equilibrium assessments. *Environ. Sci. Technol.* 26: 1955–1962.

Wernert, V., F. H. Frimmel, and P. Behra. 2003. Mercury transport through a porous medium in presence of natural organic matter. *J. Phys. IV France* 107: 1361–1364.

6

Second-Order Transport Modeling

In this chapter, an analysis is presented of a kinetic second-order approach for describing mechanisms for retention of reactive chemicals in the soil environment. The basis for this approach is that it accounts for the sites on the soil matrix that are accessible for retention of reactive chemicals in solution. The second-order approach will be incorporated into the non-equilibrium two-site model for the purpose of simulation of the potential retention during transport of reactive chemicals in soils. As will be described in a subsequent chapter, this second-order approach will be extended to the diffusion-controlled mobile-immobile (or two-region) transport model.

A main feature of the second-order model proposed here is the supposition that there exist two types of retention sites on soil matrix surfaces. Moreover, the primary difference between these two types of sites is based on the rate of the proposed kinetic retention reactions. It is also assumed that the retention mechanisms are site specific, for example, the sorbed phase on type 1 sites may be characteristically different (in their energy of reaction and/or the identity of the solute-site complex) from that on type 2 sites. An additional assumption is that the rate of solute retention reaction is a function not only of the solute concentration present in the solution phase but also of the amount of available retention sites on matrix surfaces. Another feature of the second-order approach is that an adsorption maximum (or capacity) is assumed. For a specific reactive chemical, this maximum represents the total number of adsorption sites on the soil matrix. This adsorption maximum is also considered an intrinsic property of an individual soil and is thus assumed constant (Selim and Amacher, 1988).

6.1 Second-Order Kinetics

For simplicity, we denote S_{max} to represent the total retention capacity or the maximum adsorption sites on matrix surfaces. It is also assumed that S_{max} is invariant with time. Therefore, based on the two-site approach, the total sites can be partitioned into two types such that:

$$S_{max} = (S_{max})_1 + (S_{max})_2 \tag{6.1}$$

where $(S_{max})_1$ and $(S_{max})_2$ are the total number of type 1 sites and type 2 sites, respectively. If F represents the fraction of type 1 sites to the total number of sites or the adsorption capacity for an individual soil, we thus have:

$$(S_{max})_1 = F\,S_{max} \quad \text{and} \quad (S_{max})_2 = (1-F)\,S_{max} \tag{6.2}$$

We now denote ϕ as the number of unfilled or vacant sites in the soil, such that:

$$\phi_1 = (S_{max})_1 - S_1 = F\,S_{max} - S_1 \tag{6.3}$$

$$\phi_2 = (S_{max})_2 - S_2 = (1-F)\,S_{max} - S_2 \tag{6.4}$$

where ϕ and ϕ_2 are the number of vacant sites and S_1 and S_2 are the amounts of solute retained (or the number of filled sites) on type 1 and type 2 sites, respectively. As the sites become filled or occupied by the reactive solute, the number of vacant sites approaches zero, that is, $(\phi_1 + \phi_2) \rightarrow 0$. In the meantime, the amount of solute retained by the soil matrix approaches that of the total capacity (or maximum number) of sites, $(S_1 + S_2) \rightarrow S_{max}$.

We commonly express the amount of reactive chemical retained, such as S_1 and S_2 in Equations 6.3 and 6.4 as the mass of solute per unit mass of soil (mg kg^{-1} soil). Based on the above formulations, the total number of sites S_{max}, $(S_{max})_1$, and $(S_{max})_2$ and vacant or unfilled sites ϕ_1 and ϕ_2 must also have similar dimensions. Here the units used for S and ϕ will be in terms of milligrams of solute per kilogram soil mass (mg kg^{-1} soil).

Based on this approach, reactive chemical retention mechanisms are assumed to follow a second-order kinetic reaction where the forward process is controlled by the product of the solution concentration C (mg L^{-1}) and the number of unoccupied or unfilled sites (ϕ) (Selim and Amacher, 1988). Specifically, the reactions for type 1 and type 2 sites may be expressed by the reversible processes; $t \rightarrow \infty$:

$$C + \phi_1 \underset{k_2}{\overset{k_1}{\rightleftarrows}} S_1 \tag{6.5}$$

and

$$C + \phi_2 \underset{k_4}{\overset{k_3}{\rightleftarrows}} S_2 \tag{6.6}$$

Therefore, the differential form of the kinetic rate equations for reactive chemical retention may be expressed as:

$$\rho\frac{\partial S_1}{\partial t} = k_1 \Theta\, \phi_1\, C - k_2 \rho\, S_1 \quad \text{for type 1 sites} \tag{6.7}$$

and

$$\rho \frac{\partial S_2}{\partial t} = k_3 \Theta \phi_2 C - k_4 \rho S_2 \qquad \text{for type 2 sites} \qquad (6.8)$$

where k_1 and k_2 (h^{-1}) are forward and backward rate coefficients for type 1 sites, whereas k_3 and k_4 are rate coefficients for type 2 reaction sites. In addition, Θ is the soil water content ($cm^3\ cm^{-3}$), ρ is the soil bulk density ($g\ cm^{-3}$), and t is time (h). If ϕ_1 and ϕ_2 are omitted from Equations 6.7 and 6.8, the above equations yield two first-order kinetic retention reactions (Lapidus and Amundson, 1952). However, a major disadvantage of first-order kinetic reactions is that as the concentration in solution increases, a maximum solute sorption is not attained, which implies that there is an infinite solute retention capacity of the soil or infinite number of exchange sites on the matrix surfaces. In contrast, the approach proposed here achieves maximum sorption when all unfilled sites become occupied (i.e., ϕ and $\phi_2 \to 0$).

In a fashion similar to the nonequilibrium two-site concept proposed by Selim, Davidson, and Mansell (1976), it is possible to regard type 1 sites as those where equilibrium is rapidly reached (i.e., in a few minutes or hours). In contrast, type 2 sites are highly kinetic and may require several days or months for apparent local equilibrium to be achieved. Therefore, for type 1 sites the rate coefficients k_1 and k_2 are expected to be several orders of magnitude larger than k_3 and k_4 of the type 2 sites. As $t \to \infty$, that is, when both sites achieve local equilibrium, Equations 6.7 and 6.8 yield the following expressions. For type 1 sites:

$$k_1 \Theta \phi_1 C - k_2 \rho S_1 = 0, \quad \text{or} \quad \frac{S_1}{\phi_1 C} = \frac{\Theta}{\rho} \frac{k_1}{k_2} = \omega_1 \qquad (6.9)$$

and for type 2 sites:

$$k_3 \Theta \phi_2 C - k_4 \rho S_2 = 0, \quad \text{or} \quad \frac{S_2}{\phi_2 C} = \frac{\Theta}{\rho} \frac{k_3}{k_4} = \omega_2 \qquad (6.10)$$

Here ω_1 and ω_2 represent equilibrium constants for the retention reactions associated with type 1 and type 2 sites, respectively. The formulations of Equations 6.9 and 6.10 are analogous to expressions for homovalent ion-exchange equilibrium reactions. In this sense, the equilibrium constants ω_1 and ω_2 resemble the selectivity coefficients for exchange reactions and S_{max} resembles the exchange capacity (CEC) of soil matrix surfaces. However, a major difference between ion exchange and the proposed second-order approach is that no consideration of other competing ions in solution or on matrix surfaces is incorporated into the second-order rate equations. In a strict thermodynamic sense, Equations 6.9 and 6.10 should be expressed in terms of activities rather than concentrations. However, we use the implicit assumption that solution-phase ion activity coefficients are constant in a constant ionic strength medium. Moreover, the solid-phase ion activity coefficients are assumed to be incorporated in the selectivity coefficients (ω_1 and ω_2) as in ion-exchange formulations.

Incorporation of Equations 6.1 to 6.4 into Equations 6.9 and 6.10 and further rearrangement yields the following expressions for the amounts retained by type 1 and type 2 sites at $t \rightarrow \infty$,

$$\frac{S_1}{(S_{max})_1} = \left[\frac{\omega_1 C}{1 + \omega_1 C} \right], \quad \text{and} \quad \frac{S_2}{(S_{max})_2} = \left[\frac{\omega_1 C}{1 + \omega_2 C} \right] \tag{6.11}$$

Therefore, the total amount sorbed in the soil S ($= S_1 + S_2$), is

$$\frac{S}{S_{max}} = \left[\frac{\omega_1 C}{1 + \omega_1 C} \right] F + \left[\frac{\omega_2 C}{1 + \omega_2 C} \right] (1 - F) \tag{6.12}$$

Equation 6.12 is analogous to the two-site Langmuir formulation where the amount sorbed in each region is clearly expressed. Such Langmuir formulations are commonly used to obtain independent parameter estimates for S_{max} and the affinity constants ω_1 and ω_2.

Let us now consider the case where only one type of active sites is dominant in the soil system. In a similar fashion to the formulations of Equations 6.9 and 6.10, the kinetics of the reaction can be described by the following equation (Murali and Aylmore, 1983):

$$\rho \frac{\partial S}{\partial t} = k_f \, \Theta \, \phi \, C - k_b \rho S \tag{6.13}$$

Here k_f and k_b (h^{-1}) are the forward and backward retention rate coefficients and S is the total amount of solute retained by the soil matrix surfaces. This formulation is often referred to as the kinetic Langmuir equation. In fact, Equation 6.13 at equilibrium obeys the widely recognized Langmuir isotherm equation:

$$\frac{S}{S_{max}} = \frac{\omega C}{1 + \omega C} \tag{6.14}$$

where $\omega = (\Theta k_f / \rho k_b)$ is equivalent to that of Equations 6.9 and 6.10. For a discussion of the formulation of the kinetic Langmuir equation see Rubin (1983).

It should be recognized that the unfilled or vacant sites (ϕ) in Equations 6.7, 6.8, and 6.13 are not strictly vacant. They are occupied by hydrogen, hydroxyl, or other nonspecifically adsorbed (e.g., Na, Ca, Cl, and NO_3) or specifically adsorbed (e.g., PO_4, AsO_4, and transition metals) species. Vacant or unfilled refers to sites vacant or unfilled by the specific solute species of interest. The process of occupying a vacant site by a given solute species actually is one of replacement or exchange of one species for another. However, the simplifying assumption on which this model is based is that the filling of sites by a particular solute species need not consider the corresponding replacement of species already occupying the sites. The Langmuir-type approach considered here (Equations 6.11 to 6.14) is a specialized case of an ion-exchange formulation (ElPrince and Sposito, 1981). Alternatively, the competitive Langmuir approach may be used if the identities of the replaced solute species are known.

6.1.1 Transport Model

Incorporation of the second-order reactions into the classical (convection-dispersion) transport equation yields (Brenner, 1962; Nielsen, van Genuchten, and Biggar, 1986):

$$\Theta \frac{\partial C}{\partial t} + \rho \left(\frac{\partial S_1}{\partial t} + \frac{\partial S_2}{\partial t} \right) = \Theta D \frac{\partial^2 C}{\partial x^2} - q \frac{\partial C}{\partial x} - Q \tag{6.15}$$

where D is the hydrodynamic dispersion coefficient (cm^2 h^{-1}), q is Darcy's water flux (cm h^{-1}), and x is depth (cm). Here the term Q is a sink representing the rate of irreversible reactive chemical reactions by direct removal from the soil solution (mg h^{-1} cm^{-3}). In this model, the sink term was expressed in terms of a first-order irreversible reaction for reductive sorption, precipitation, or internal diffusion:

$$Q = \Theta k_s C \tag{6.16}$$

where k_s is the rate constant for the irreversible reaction (h^{-1}). Equation 6.16 is similar to that for a diffusion-controlled precipitation reaction if one assumes that the equilibrium concentration for precipitation is negligible and that k_s is related to the diffusion coefficient.

For convenience, we define the dimensionless variables:

$$X = \frac{x}{L} \tag{6.17}$$

$$T = \frac{q\,t}{L\,\Theta} \tag{6.18}$$

$$c = \frac{C}{C_o} \tag{6.19}$$

$$s = \frac{S}{S_{max}} \tag{6.20}$$

$$\Phi = \frac{\varphi}{S_{max}} \tag{6.21}$$

$$P = \frac{qL}{D\Theta} \tag{6.22}$$

where T is dimensionless time equivalent to the number of pore volumes leached through a soil column of length L, and P is the Peclet number (Brenner, 1962). Given the above variables, Equations 6.15, 6.7, and 6.8 are rewritten in dimensionless form, respectively, as (Selim and Amacher, 1988):

$$\frac{\partial c}{\partial T} + \Omega \left(\frac{\partial s_1}{\partial T} + \frac{\partial s_2}{\partial T} \right) = \frac{1}{P} \frac{\partial^2 c}{\partial X^2} - \frac{\partial c}{\partial X} - \kappa_s c \tag{6.23}$$

$$\frac{\partial s_1}{\partial T} = \kappa_1 \Phi_1 c - \kappa_2 s_1 \tag{6.24}$$

$$\frac{\partial s_2}{\partial T} = \kappa_3 \Phi_2 c - \kappa_4 s_2 \tag{6.25}$$

where

$$\Omega = \frac{S_{max} \rho}{C_o \Theta} \tag{6.26}$$

$$\kappa_s = \frac{k_s \Theta L}{q} \tag{6.27}$$

$$\kappa_1 = \frac{k_1 \Theta^2 C_o L}{\rho q} \quad \text{and} \quad \kappa_3 = \frac{k_3 \Theta^2 C_o L}{\rho q} \tag{6.28}$$

Here, κ_s, κ_1, κ_2, κ_3, and κ_4 are dimensionless kinetic rate coefficients that incorporate q and L. As will be shown in a later section, these dimensionless variables (including Ω, c, s, and ϕ) represent a convenient way to study the sensitivity of the model to reduced variables.

For the purpose of simulation and model evaluation, the appropriate initial and boundary conditions associated with Equations 6.23 to 6.25 were as follows. We chose uniform initial solute concentration C_i in a finite soil column of length L such that:

$$c = c_i \quad (T = 0 \quad \text{and} \quad 0 < X < 1) \tag{6.29}$$

We also assume that an input solute solution pulse having a concentration C_o was applied at the soil surface for a (dimensionless) time T_p and was then followed by a solute-free solution. As a result, at the soil surface, the following third-type boundary conditions were used (Selim and Mansell, 1976):

$$1 = c - \frac{1}{P} \frac{\partial c}{\partial X} \quad (X = 0, T < T_p) \tag{6.30}$$

$$0 = c - \frac{1}{p} \frac{\partial c}{\partial X} \quad (X = 0, T > T_p) \tag{6.31}$$

At $x = L$, we have:

$$\frac{\partial c}{\partial X} = 0 \quad (X = 1, T > 0) \tag{6.32}$$

The differential equations of the second-order model described above are of the nonlinear type and analytical solutions are not available. Therefore, Equations 6.21 to 6.23 must be solved numerically. A finite-difference approximation (explicit-implicit) subject to the above initial and boundary conditions can be derived as was carried out by Selim and Amacher (1988) and documented in Selim, Amacher, and Iskandar (1990).

6.2 Sensitivity Analysis

Several simulations were performed to illustrate the kinetic behavior of solute retention as governed by the proposed second-order reaction. We assumed a no-flow condition to describe time-dependent batch (sorption-desorption) experiments. The problem becomes an initial-value problem where closed-form solutions are available.

The retention results shown in Figure 6.1 illustrate the influence of the rate coefficients (k_1 and k_2) on the shape of sorption isotherms (S versus C). The parameters chosen were those of a soil initially devoid of solute

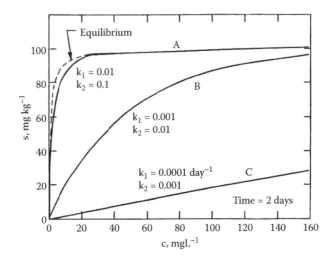

FIGURE 6.1
Effect of rate coefficients on sorption isotherms using the second-order kinetic model. (From H. M. Selim and M. C. Amacher. 1997. *Reactivity and Transport of Heavy Metals in Soils.* Boca Raton, FL: CRC Press. With permission.)

($C_i = S_i = 0$ at $t = 0$) and a soil-to-solution ratio (ρ/Θ) of 1:10, which is commonly used in batch experiments. Since the amount sorbed was assumed to be initially zero, larger values for k_2 than k_1 were selected in our simulations to induce reverse (desorption) reactions.

As shown in Figure 6.1, after 2 days of reaction, isotherm A, where k_1 and k_2 were 0.01 and 0.1 day^{-1}, respectively, appears closer to the equilibrium isotherm than other cases shown. The equilibrium case was calculated using Equation 6.9 and represents an isotherm at $t \to 4$ for a soil having values of k_1 and k_2 that are extremely large. Isotherms B and C represent cases where both k_1 and k_2 values were reduced in comparison to those for isotherm A, by one and two orders of magnitude, respectively. For both cases, the isotherms deviate significantly from the equilibrium case. It is apparent from curve **C** that 2 days of reaction is insufficient to attain equilibrium and a sorption maximum is not apparent from the shape of the isotherm. Moreover, it is perhaps possible to consider a linear-type isotherm for the concentration range shown. However, as much as 100 days or more of reaction time is necessary to achieve apparent equilibrium conditions. This is illustrated in Figure 6.2, where the influence of time of reaction using the second-order model is shown.

The influence of the sorption maxima (S_{max}) on the retention isotherms is shown in Figure 6.3. The parameters selected were similar to those of Figure 6.2 except that a contact time of 10 days was chosen. As expected, the isotherms reached their respective maxima at lower C values with decreasing S_{max}. The results of Figure 6.3 also indicate a steep gradient of the retention

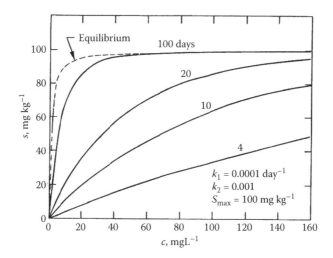

FIGURE 6.2
Effect of time of retention on sorption isotherms using the second-order kinetic model. (From H. M. Selim and M. C. Amacher. 1997. *Reactivity and Transport of Heavy Metals in Soils.* Boca Raton, FL: CRC Press. With permission.)

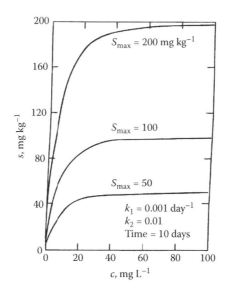

FIGURE 6.3
Effect of total number of sites (S_{max}) on the shape of sorption isotherms using the second-order kinetic model. (From H. M. Selim and M. C. Amacher. 1997. *Reactivity and Transport of Heavy Metals in Soils.* Boca Raton, FL: CRC Press. With permission.)

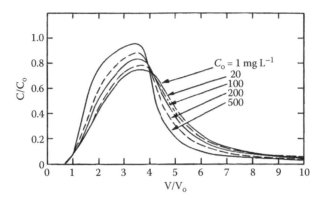

FIGURE 6.4
Effluent concentration distributions for different initial concentrations (C_o) using the second-order model. (From H. M. Selim and M. C. Amacher. 1997. *Reactivity and Transport of Heavy Metals in Soils*. Boca Raton, FL: CRC Press. With permission.)

isotherms in the low concentration range. Such a retention behavior has been observed by several scientists for a number of reactive solutes. The simulations also illustrate clearly the influence of the sorption maxima on the overall shape of the isotherms. The influence of other parameters such as F, k_3, and k_4 on retention kinetics can be easily deduced and is thus not shown.

Figures 6.4, 6.5, and 6.6 are selected simulations that illustrate the transport of a reactive solute with the second-order model as the governing retention mechanism. The parameters selected for the sensitivity analysis were $\rho = 1.25$ g cm^{-3}, $\Theta = 0.4$ cm^3 cm^{-3}, $L = 10$ cm, $C_i = 0$, $C_o = 10$ mg L^{-1}, $F = 0.50$, and $S_{max} = 200$ mg kg^{-1}. Here we assumed a solute pulse was applied to a

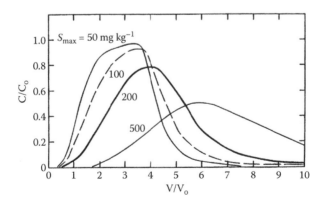

FIGURE 6.5
Effluent concentration distributions for different S_{max} values using the second-order model. (From H. M. Selim and M. C. Amacher. 1997. *Reactivity and Transport of Heavy Metals in Soils*. Boca Raton, FL: CRC Press. With permission.)

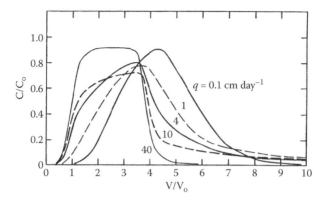

FIGURE 6.6
Effluent concentration distributions for different flux (v) values using the second-order model. (From H. M. Selim and M. C. Amacher. 1997. *Reactivity and Transport of Heavy Metals in Soils.* Boca Raton, FL: CRC Press. With permission.)

fully water-saturated soil column initially devoid of a particular reactive chemical of interest. In addition, a steady water-flow velocity (q) was maintained constant with a Peclet number P (= $q/\Theta D$) of 25. The length of the pulse was assumed to be three pore volumes, which was then followed by several pore volumes of a reactive chemical-free solution. The rate coefficients selected were 0.01, 0.1, 0.001, and 0.01 day^{-1} for k_1, k_2, k_3, and k_4, respectively. As a result, the equilibrium constants ω_1 and ω_2 for sites 1 and 2, respectively, were identical.

Figure 6.4 shows breakthrough curves (BTCs), which represent the relative effluent concentration (C/C_0) versus effluent pore volume (V/V_0), for several input C_0 values. The shape of the BTCs is influenced by the input solute concentration and is due to the nonlinearity of the proposed second-order retention mechanism. The simulated results also indicate that for high C_0 values the BTCs appear less retarded and have sharp gradients on the desorption (or right) side. In contrast, for low C_0 values the general shape of the BTCs appear to be kinetic in nature. Specifically, as C_0 decreases, a decrease in maximum or peak concentrations and extensive tailing of the desorption side of the BTCs can be observed.

The influence of the total number of (active) sites (S_{max}) on the BTC is clearly illustrated by the cases given in Figure 6.5. Here the value of C_0 was chosen constant ($C_0 = 10$ mg L^{-1}). The BTCs show that an order of magnitude increase in S_{max} (from 50 to 500 mg $\cong$ kg^{-1}) resulted in an approximately three pore volume shift in peak position. In addition, for high S_{max} values extensive tailing and an overall decrease of effluent concentrations (C/C_0) were observed.

The influence of the flow velocity (q) on the shape of the BTC is somewhat similar to that of the rate coefficients for retention provided that the Peclet number (P) remains constant. This is illustrated by the simulations shown in Figure 6.6 for a wide range of flow velocities. For $q = 40$ cm day^{-1}, the retention reactions associated with type 1 sites were not only dominant but also closer to local equilibrium than those for type 2 sites (results not shown). This is a direct consequence of the limited solute residence time encountered when the fluid flow velocity is exceedingly high. Type 2 sites, which may be considered highly kinetic, were far removed from equilibrium such that only a limited amount of solute was retained from the soil solution. Under such conditions, the number of available sites (ϕ) remains high and the retention capacity of the soil matrix is therefore not achieved. In fact, for $q = 40$ cm/day, the BTC describes closely a one site retention mechanism as indicated by the low retardation and lack of tailing of the desorption side. As the flow velocity decreases the solute residence time increases and more time is available for the highly kinetic type 2 sites to retain solute species from the soil solution. In addition, for extremely small velocities the BTC should indicate maximum solute retention during transport. This probably resembles the BTCs with $q = 0.1$ cm/day, which indicates the highest solute retardation shown. For intermediate velocities (q from 1 to 10), however, the respective BTCs indicate relatively moderate degrees of retardation as well as tailing, which is indicative of kinetic retention mechanisms.

In the BTCs shown in Figures 6.4, 6.5, and 6.6 the irreversible retention mechanism for reactive chemical removal (via the sink term) was ignored. The influence of the irreversible kinetic reaction (e.g., precipitation) is a straightforward one as shown in Figure 6.7. This is manifested by the lowering of solute concentration for the overall BTC for increasing values of k_s. Since a first-order reaction was assumed the lowering of the BTC is proportional to the solution concentration.

In previous BTCs, the sensitivity of model predictions (output) of the second-order approach to selected parameters was discussed. It is convenient, however, to assess model sensitivity using dimensionless parameters

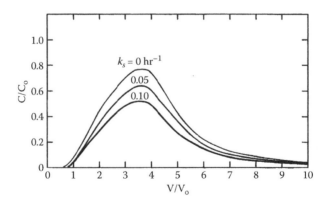

FIGURE 6.7
Effluent concentration distributions for values of the irreversible rate coefficient (k_s) using the second-order model. (From H. M. Selim and M. C. Amacher. 1997. *Reactivity and Transport of Heavy Metals in Soils*. Boca Raton, FL: CRC Press. With permission.)

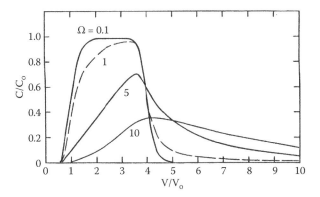

FIGURE 6.8
Effluent concentration distributions for different values of the parameter Ω of the second-order model. (From H. M. Selim and M. C. Amacher. 1997. *Reactivity and Transport of Heavy Metals in Soils*. Boca Raton, FL: CRC Press. With permission.)

such as those defined by Equations 6.17 to 6.22. The use of dimensionless parameters offers a distinct advantage over the use of conventional parameters since they provide a wide range of application as well as further insight on predictive behavior of the model. Figures 6.8 to 6.10 are simulations that illustrate the transport of a reactive solute with the second-order model for selected dimensionless parameters. Unless otherwise indicated the values for the dimensionless parameters Ω, κ_1, κ_2, κ_3, κ_4, F, κ_s, P, and T_p were 5, 1, 1, 0.1, 0.1, 0, 0.5, 25, and 1, respectively.

Figure 6.8 shows BTCs of a reactive solute for several values of Ω. The figure indicates that the shape of the BTCs is influenced drastically by the value

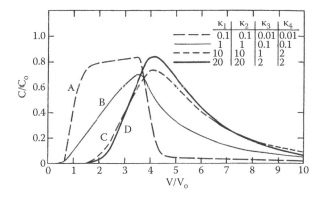

FIGURE 6.9
Effluent concentration distributions for different values of rate coefficients (κ_1, κ_2, κ_3, and κ_4) using the second-order model. (From H. M. Selim and M. C. Amacher. 1997. *Reactivity and Transport of Heavy Metals in Soils*. Boca Raton, FL: CRC Press. With permission.)

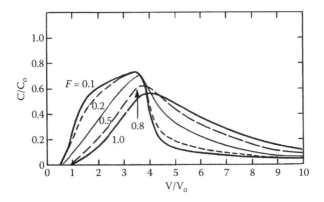

FIGURE 6.10
Effluent concentration distributions for different values of the fraction of sites F using the second-order model. (From H. M. Selim and M. C. Amacher. 1997. *Reactivity and Transport of Heavy Metals in Soils*. Boca Raton, FL: CRC Press. With permission.)

of Ω. This is largely due to the nonlinearity of the proposed second-order retention mechanism. As given by Equation 6.26, Ω represents the ratio of total sites (S_{max}) to input (pulse) solute concentration (C_0). Therefore, for small values of Ω (e.g., $\Omega = 0.1$), the simulated BTC is very similar to that for a nonretarded solute due to the limited number of sites (S_{max}) in comparison to C_0. In contrast, large values of Ω resulted in BTCs that indicate increased retention as manifested by the right shift of peak concentration of the BTCs. In addition, for high Ω, extensive tailing as well as an overall decrease of effluent concentration were observed.

The effect of the dimensionless reaction rate coefficients (κ_1, κ_2, κ_3, and κ_4) of the two-site model on solute retention and transport is illustrated by the BTCs of Figure 6.9 where a range of rate coefficients differing by three orders of magnitude were chosen. For the BTCs shown, the rate coefficients for type 2 sites were chosen to be one order of magnitude smaller than those associated with type 1 sites. These BTCs indicate that, depending on the values of κ_1, κ_2, κ_3, and κ_4, two extreme cases can be illustrated. For large values of κ, rapid sorption-desorption reactions occurred for both type 1 and type 2 sites. Rapid reactions indicate that the retention process is less kinetic and BTCs can approximate local equilibrium conditions in a relatively short contact time. Examples are those of curves C and D. In contrast, for extremely small values of κ_1, κ_2, κ_3, and κ_4 (or small residence time), little retention takes place and the shape of the BTC resembles that for a nonreactive solute (see curve A). The behavior of all illustrated BTCs is consistent with those for first-order kinetic and for two-site nonlinear equilibrium-kinetic reactions.

Figure 6.10 shows BTCs for several values of the fraction of sites parameter F. There are similar features between the BTCs of Figures 6.8 and 6.9 and those illustrated in Figure 6.10. For F = 1, all the sites are type 1 sites which we designated earlier as those sites of strong kinetic influence due to their large values of κ_1 and κ_2. As the contribution of type 2 sites increases (or F decreases), the shape of the BTCs becomes increasingly less kinetic with significant decreases in the amount of solute retention.

6.3 Experimental Data on Retention

The second-order model (SOM) was applied by Selim and Amacher (1988) and Amacher and Selim (1994) to describe batch and miscible displacement data sets for Cr(VI) retention and transport in three different soils. Furthermore, they attempted, whenever possible, to utilize parameters that were either independently measured or estimated by indirect means. The parameters that were estimated included the adsorption maximum S_{max}, the fraction of sites F, and the kinetic rate coefficients associated with the reversible and irreversible mechanisms.

Retention data sets for Cr(VI) by three soils (Cecil, Olivier, and Windsor soils) after 336 h of reaction (Figure 6.11) were used to arrive at independent estimates of S_{max} and F. Specifically, the two-site Langmuir equation (6.12) was used to describe Cr(VI) retention results using a nonlinear, least-squares, parameter-optimization scheme. It was assumed that the reactions between Cr(VI) in solution and the two types of sites had attained equilibrium in 336 h even though small amounts of Cr(VI) were still being retained by the soil. The continuing

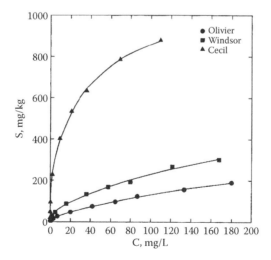

FIGURE 6.11
Two-site Langmuir sorption curves for Cr(VI) retention by Olivier, Windsor, and Cecil soils after 336 h of reaction. (From H. M. Selim and M. C. Amacher. 1997. *Reactivity and Transport of Heavy Metals in Soils*. Boca Raton, FL: CRC Press. With permission.)

reaction between Cr(VI) in solution and the soil was ascribed to an irreversible reaction, which is included in the model (Equation 6.15). It is important to realize that only the reactions of Cr(VI) with the two types of reaction sites were assumed to attain equilibrium in 336 h. Overall retention had not reached equilibrium because of the irreversible reaction. However, if the magnitude of the irreversible term is small as is the case here, then reliable estimates of S_{max} and F can be made, although the actual S_{max} is somewhat smaller. The statistical results indicated a close approximation of the two-site Langmuir equation to the experimental sorption isotherms shown in Figure 6.11.

Data sets of time-dependent retention of Cr(VI) by Olivier, Windsor, and Cecil soils and for several input (initial) concentrations (C_o) were used to provide estimates of the kinetic rate coefficients of the second-order approach. Both S_{max} and F values previously obtained were used as model inputs and nonlinear parameter optimization was used to estimate k_1, k_2, k_3, k_4, and k_s. Two versions of SOM were used: a three-parameter or a one-site version (k_1, k_2, and k_s) in which S_{max} was not differentiated into type 1 and type 2 sites ($F = 1$) and a five-parameter or a two-site version (k_1, k_2, k_3, k_4, and k_s) in which two types of reaction sites were considered. For most values of C_o, either the three- or five-parameter versions described the data adequately, with high R^2 values and low parameter standard errors. The exception was the description of retention data for Olivier soil at high C_o where the retention of Cr(VI) was not highly kinetic and more scatter in experimental data was observed. Model calculations and data for the Olivier, Windsor, and Cecil soils are shown in Figures 6.12, 6.13, and 6.14, respectively.

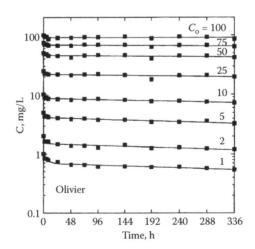

FIGURE 6.12
Time-dependent retention of Cr(VI) by Olivier soil. Closed squares are the data points and solid lines are second-order model predictions for different initial concentration curves ($C_o = 1$, 2, 5, 10, 25, 50, 75, and 100 mg @ L⁻¹). (From H. M. Selim and M. C. Amacher. 1997. *Reactivity and Transport of Heavy Metals in Soils*. Boca Raton, FL: CRC Press. With permission.)

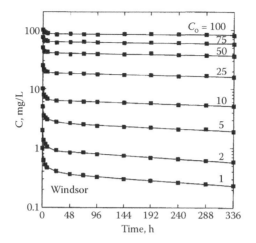

FIGURE 6.13

Time-dependent retention of Cr(VI) by Windsor soil. Closed squares are the data points and solid lines are second-order model predictions for different initial concentration curves ($C_o = 1$, 2, 5, 10, 25, 50, 75, and 100 mg @ L^{-1}). (From H. M. Selim and M. C. Amacher. 1997. *Reactivity and Transport of Heavy Metals in Soils*. Boca Raton, FL: CRC Press. With permission.)

If the fraction of type 1 sites is small as was the case with the Olivier and Windsor soils, then their contribution to the kinetic solute retention curve will be small and indistinguishable at high solute concentrations. For the Cecil soil where the fraction of type 1 sites was significant ($F = 0.224$), the five-parameter model version was superior to the three-parameter version

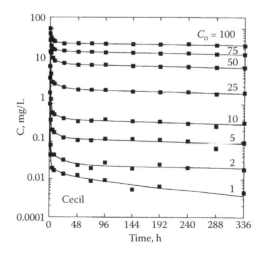

FIGURE 6.14

Time-dependent retention of Cr(VI) by Cecil soil. Closed squares are data points and solid lines are second-order model predictions for different initial concentration curves ($C_o = 1$, 2, 5, 10, 25, 50, 75, and 100 mg @ L^{-1}). (From H. M. Selim and M. C. Amacher. 1997. *Reactivity and Transport of Heavy Metals in Soils*. Boca Raton, FL: CRC Press. With permission.)

at all values of C_0 except for $C_0 = 1$ mg L^{-1}. Therefore, the applicability of the five-parameter version to a wide range of solute concentrations was directly related to the magnitude of the fraction of type 1 sites. As F increased (Olivier < Windsor < Cecil), the concentration range over which the five-parameter version provided a better description of the data than the three-parameter version increased. The shapes of the experimental curves and model calculations (Figures 6.12, 6.13, and 6.14) are influenced by C_0. At higher C_0, retention of Cr(VI) from solution was far less kinetic than at lower C_0. This behavior is as expected if the concentration of one or more reaction sites limits reaction rates. At $C_0 = 100$ mg L^{-1} there were 4 mg of Cr(VI) available for reaction in the 40 mL of solution volume used in the experiment. The maximum possible amounts of Cr(VI) that could be sorbed by the 4 g of each soil used in the experiment were 1.9, 2.9, and 4.5 mg (solute weight basis) for Olivier, Windsor, and Cecil soils, respectively. Thus, maximum possible Cr(VI) retention in the Cecil soil was about equal to the amount of Cr(VI) available for retention, but was much less in the Olivier and Windsor soils than the amount of Cr available. Since the number of type 1 sites was much less than the total, their contribution to the overall reaction is actually quite negligible at high solute concentrations. The influence of S_{max}/C_0 or Ω ($\Omega = \rho S_{max}/C_0 \Theta$) on solute retention during transport is illustrated in Figure 6.8.

In general the SOM approach well described the data sets shown in Figures 6.12, 6.13, and 6.14. Moreover, the retention processes responsible for Cr(VI) retention may include physical adsorption, formation of outer- or inner-sphere surface complexes, ion exchange, surface precipitation, etc. Furthermore, subsequent solute transformations on the soil surface or internal diffusion into soil particles may occur.

6.4 Experimental Data on Transport

Chromium BTCs from the miscible displacement experiments for all three soils are shown in Figures 6.15, 6.16, and 6.17. For Cecil and Windsor soils the measured BTCs appear to be highly kinetic, with extensive tailing. For Olivier soil little tailing was observed and approximately 100% of the applied Cr(VI) pulse was recovered. These results are consistent with the batch data where the irreversible reaction parameter (k_s) was found to be quite small.

In order to examine the capability of the SOM, the necessary model parameters must be provided. Values for the dispersion coefficients (D) were obtained from BTCs of tracer data for ^3H_2O and ^{36}Cl. Other model parameters such as ρ, Θ, and q were measured for each soil column. In addition, values for S_{max} and F used in describing Cr (VI) BTCs using the SOM were those derived from adsorption isotherms shown in Figure 6.11. Direct measurement of these parameters by other than parameter-optimization techniques is not available. Moreover,

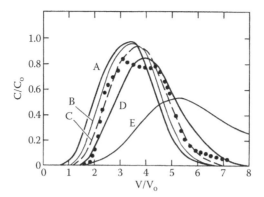

FIGURE 6.15
Effluent concentration distributions for Cr(VI) in Olivier soil. Curves A, B, C, D, and E are predictions using the second-order model with batch rate coefficients for C_0 of 100, 25, 10, 5, and 1 mg/L, respectively (From H. M. Selim and M. C. Amacher. 1997. *Reactivity and Transport of Heavy Metals in Soils*. Boca Raton, FL: CRC Press. With permission.)

we utilized the reaction rate coefficients k_1, k_2, k_3, k_4, and k_s as obtained from the batch kinetic data in the predictions of Cr(VI) BTCs. In the following discussion, predicted curves imply the use of independently measured parameters in model calculations as was carried out here using the batch-derived parameters.

The predicted BTCs shown in Figures 6.15, 6.16, and 6.17 were obtained using different sets of rate coefficients (k_1, k_2, k_3, k_4, and k_s) in the SOM. This

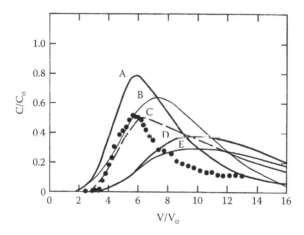

FIGURE 6.16
Effluent concentration distributions for Cr(VI) in Windsor soil. Curves A, B, C, D, and E are predictions using the second-order model with batch rate coefficients for C_0 of 25, 10, 5, 2, and 1 mg/L, respectively. (From H. M. Selim and M. C. Amacher. 1997. *Reactivity and Transport of Heavy Metals in Soils*. Boca Raton, FL: CRC Press. With permission.)

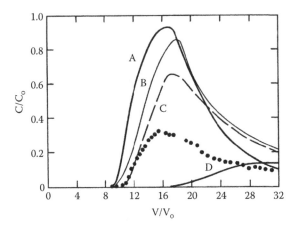

FIGURE 6.17
Effluent concentration distributions for Cr(VI) in Cecil soil. Curves A, B, C, and D are predictions using the second-order model with batch rate coefficients for C_o of 100, 50, 25, and 10 mg/L, respectively. (From H. M. Selim, and M. C. Amacher. 1997. *Reactivity and Transport of Heavy Metals in Soils.* Boca Raton, FL: CRC Press. With permission.)

is because no single or unique set of values for these rate coefficients was obtained from the batch data, rather a dependence of rate coefficients on input concentration (C_o) was observed. For all soils, several features of the predicted BTCs are similar and indicate dependence on the rate coefficients used in model calculations. Increased sorption during transport, lowering of peak concentrations, and increased tailing were predicted when batch rate coefficients from low initial concentrations (C_o) were used.

Figures 6.15 to 6.17 shows that using batch rate coefficients at $C_o = 100$ mg L^{-1}, which is the concentration of Cr(VI) in the input pulse, overestimated Cr(VI) retention. Reasons for this overestimation, which has been observed by other scientists, are not fully understood. Rate coefficients based on batch experiments varied with C_o, which would be expected for pseudo rate coefficients. Unless the concentrations of unaccounted for reaction components remain relatively constant over the course of the experiment, rate coefficients will vary with C_o because they implicitly include concentrations of other reaction components. Much larger changes were observed in Cr(VI) concentrations in column effluent (pulse input) than in the batch solutions. Moreover, in the batch experiment reaction products are not removed and the reaction is considered a closed system. In the column experiment, on the other hand, solutes are continually displaced. Also, in batch experiments the soil suspension is continuously shaken, whereas in the miscible displacement column experiment fluid flow is dominant and no such agitation of the solid phase occurs. Model parameters k_1, k_2, k_3, and k_4 did not vary over the concentration range used in the batch experiment. Thus, a valid set of rate coefficients from the batch experiment is readily available to cover the range of concentrations found in the miscible displacement experiment. The only exception is for the irreversible parameter k_s, which increased as the input concentration (C_o) increased.

For Olivier soil, predicted BTCs using the SOM indicate that the use of batch rate coefficients from either C_o of 10 or 25 mg L^{-1} provided surprisingly good overall descriptions of the experimental results (Figure 6.15). Less than adequate predictions were obtained for the highly kinetic Cecil and Windsor soils, however. In fact, no one set of batch rate coefficients successfully described both the adsorption and the desorption sides of Windsor or Cecil BTCs. For both soils, closest predictions were realized using batch rate coefficients from C_o of 10 or 25 mg L^{-1}. This is a similar finding to that based on the predictions for Olivier soil.

6.4.1 Modified Second-Order Approach

A major modification of the second-order model (SOM) approach was introduced by Ma and Selim (1994, 1998) where few model parameters are required. The primary difference of the modified approach from that of the original second-order formulation presented earlier is the assumption that vacant sites are equally accessible and can thus be occupied by either S_1 or S_2. That is, S_1 and S_2 can compete for the unoccupied adsorption sites regardless of whether they are type 1 or type 2 sites. Therefore, it is assumed that adsorption sites are related and affected by each other and adsorption of one species may block the adsorption sites of the other type. In other words, the total adsorption sites (S_{max}) were not partitioned between S_1 and S_2 phases based on a fraction of sites F. Instead, it is now assumed that the vacant sites are available to both types of sites and F is no longer required and the amount of solute adsorbed on each type of sites is only determined by the rate coefficients associated with each type of sites. As a result, sites associated with fast reactions will compete for available sites before slow sites are filled. Such mechanisms are in line with observations where rapid (equilibrium type) sorption is first encountered followed by slow types of retention reactions. Therefore, based on the above, we defined ϕ (µg g^{-1}) as the number of vacant sites, which is dependent on the number of sites occupied by S_2 and S_2 such that:

$$\phi = S_{max} - S_1 - S_2 \tag{6.33}$$

and governing kinetic expressions for the rate of reactions for solutes present in the soil solution phase (C) for the type 1 and type 2 sites may be written as:

$$\rho \frac{\partial S_1}{\partial t} = k_1 \Theta \phi C - k_2 \rho S_1 \quad \textit{for type 1 sites} \tag{6.34}$$

$$\rho \frac{\partial S_2}{\partial t} = k_3 \Theta \phi C - k_4 \rho S_2 \quad \textit{for type 2 sites} \tag{6.35}$$

At large time ($t \rightarrow \infty$) when both sites achieve local equilibrium yield the following expressions hold. For type 1 sites:

$$k_1 \Theta \phi C - k_2 \rho S_1 = 0, \quad or \quad \frac{S_1}{\phi C} = \frac{\Theta}{\rho} \frac{k_1}{k_2} = \omega_1 \tag{6.36}$$

and for type 2 sites:

$$k_3 \Theta \phi C - k_4 \rho S_2 = 0, \quad or \quad \frac{S_2}{\phi C} = \frac{\Theta}{\rho} \frac{k_3}{k_4} = \omega_2 \tag{6.37}$$

Here ω_1 and ω_2 represent equilibrium constants for the retention reactions associated with type 1 and type 2 sites, respectively. These formulations are analogous to expressions for the original second-order formulation, excepting that ω_1 and ω_2 are functions of the vacant sites ϕ.

In the following analysis we followed similar overall structure for the second-order formulation to that described earlier where three types of retention sites are considered with one equilibrium-type site (S_e) and two kinetic-type sites, S_1 and S_2. Therefore, we have ϕ now related to the sorption capacity (S_{max}) by:

$$S_{max} = \phi + S_e + S_1 + S_2 \tag{6.38}$$

The governing retention reactions can be expressed as (Ma and Selim, 1998):

$$\rho S_e = K_d \Theta C \phi \tag{6.39}$$

$$\rho \frac{\partial S_1}{\partial t} = k_1 \Theta \phi C - k_2 \rho S_1 \tag{6.40}$$

$$\rho \frac{\partial S_2}{\partial t} = k_3 \Theta \phi C - \rho(k_4 + k_4)S_2 \tag{6.41}$$

$$\rho \frac{\partial S_{irr}}{\partial t} = k_s \Theta C \tag{6.42}$$

The unit for K_e is $cm^3 \, \mu g^{-1}$, k_1 and k_3 have a derived unit of $cm^3 \, \mu g^{-1} h^{-1}$; k_2, k_4, k_5, and k_s are assigned a unit of h^{-1}.

6.5 Experimental Data on Retention

The input parameter S_{max} of the second-order model is a major parameter and represents the total sorption of sites. S_{max}, which is often used to characterize reactive chemical sorption, can be quite misleading if the experimental data do not cover a sufficient range of solution concentration and if other conditions such as the amounts initially sorbed prevail (Houng and Lee, 1998). In an arsenic adsorption study, Selim and Zhang (2007) used S_{max} in the SOM based on average values as determined from the Langmuir isotherm equation. This is a simple approach to obtain S_{max} estimated when a direct measurement of the sorption capacity is not available. Figures 6.18 to 6.21 show simulated adsorption results using the SOM based on two model versions for three different soils. Based on visual observations of the overall fit of the model to the experimental data, the

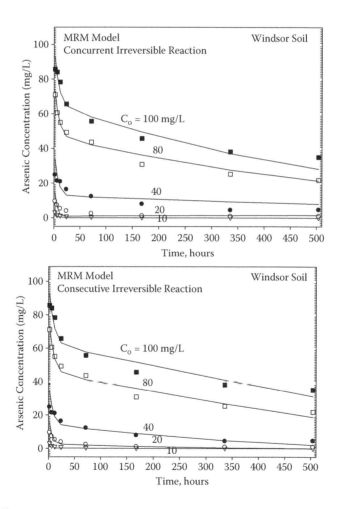

FIGURE 6.18
Experimental results of As(V) concentration in soil solution for Windsor soil versus time for all values of C_o. Dashed curves were obtained using the multireaction model with concurrent (top figure) and consecutive (bottom figure) irreversible reactions.

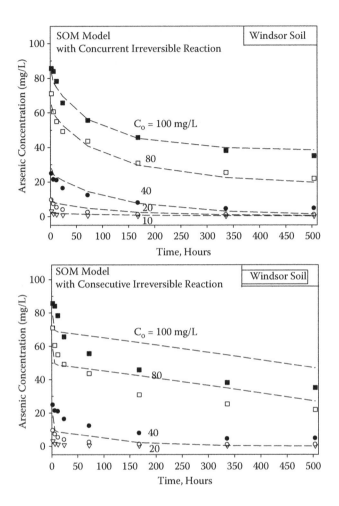

FIGURE 6.19
Experimental results of As(V) concentration in soil solution for Windsor soil versus time for all values of C_o. Dashed curves were obtained using the second-order model with concurrent (top figure) and consecutive (bottom figure) irreversible reactions.

SOM provided good overall predictions of the kinetic adsorption data for arsenic.

The question arises whether SOM model improvements can be realized when one relaxes the assumption of the use of Langmuir S_{max} and utilizes parameter optimization to arrive at a best estimate of the rate coefficients (e.g., k_1, k_2, and k_{irr} or k_3, k_4, and k_5) as well as S_{max}. Based on these results, Selim and Zhang (2007) concluded that the use of Langmuir S_{max} as an input parameter provided good predictions of the adsorption results. Moreover, retention kinetics predictions of As(V) shown are in agreement with the biphasic

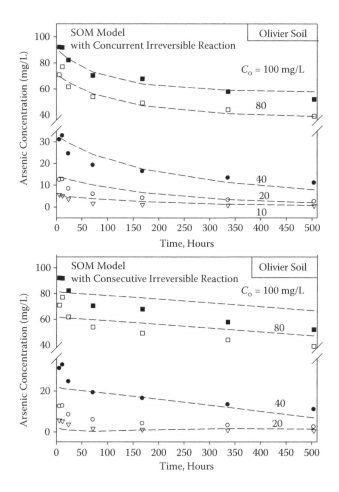

FIGURE 6.20

Experimental results of As(V) concentration in soil solution for Olivier soil versus time for all values of C_o. Dashed curves were obtained using the second-order model with concurrent (top figure) and consecutive (bottom figure) irreversible reactions.

arsenic adsorption behavior observed on several soil minerals (Fuller et al., 1993; Raven, Jain, and Loeppert, 1998; Arai and Sparks, 2002) as well as whole soils (Elkhatib, Bennett, and Wright, 1984; Carbonell-Barrachina et al., 1996) over different time scales (minutes to months).

A comparison of the multireaction (MRM) and second-order (SOM) models for their capability to predict arsenic concentration with time is given in Figure 6.22. Selim and Zhang (2007) found that several model versions fit the data equally well, but the sorption kinetics prediction capability varied among the soils investigated. MRM was superior to SOM and the use of irreversible reactions for the model formulations was essential. They also found

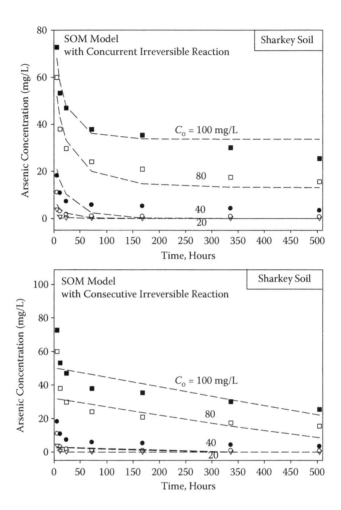

FIGURE 6.21
Experimental results of As(V) concentration in soil solution for Sharkey soil versus time for all values of C_o. Dashed curves were obtained using the second-order model with concurrent (top figure) and consecutive (bottom figure) irreversible reactions.

that incorporation of an equilibrium sorbed phase into the various model versions for As(V) predictions should be avoided.

The success of the SOM approach in describing As(v) retention results is significant since the SOM formulation described in this chapter has not been applied to metalloid elements like arsenic. Previous use of the SOM formulation, which included a partitioning of the sites, indicated that for Cr and Zn the rate coefficients were highly concentration dependent (Selim and Amacher, 1988; Hinz, Buchter, and Selim, 1992). Selim and Ma (2001)

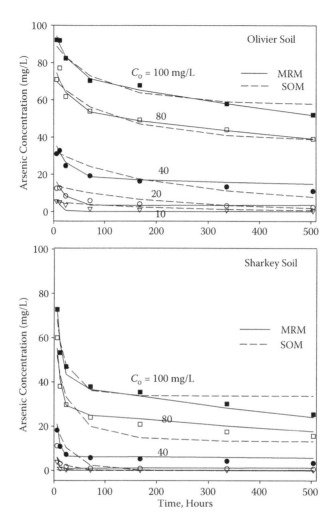

FIGURE 6.22
Experimental and predicted As(V) concentration versus time for Olivier and Sharkey soil and several input (initial) concentrations (C_o). Predictions were obtained using the multireaction and second-order models.

successfully utilized the SOM model to describe Cu adsorption as well as desorption or release following sorption. They concluded that the use of consecutive irreversible reactions (k_s in Figure 6.1) provided improvements in the description of the kinetic sorption and desorption of Cu compared to the concurrent irreversible reaction (k_{irr}). This finding is contrary to that from this study for arsenic adsorption for all three soils. Such contradictions are not easily explained and are thus a subject for future research.

6.6 Experimental Data on Transport

In the first example, Zhang and Selim (2011) investigated the capability of SOM in describing the mobility and reactivity of arsenite in two different soils. Results from their miscible displacement experiments indicated that for both soils, arsenite BTCs exhibited strong retardation, with diffusive effluent fronts followed by slow release or tailing during leaching. They also found that SOM, which accounts for equilibrium, reversible, and irreversible retention mechanisms, well described arsenite transport results from the soil columns. Figures 6.23 and 6.24 are examples of predictions based on several versions of the SOM for Windsor and Olivier soils. Based on these results, Zhang and Selim (2011) argued that the transport patterns of arsenite in these soils are indicative of dominance of kinetic retention during transport in soils. Zhang and Selim (2011) concluded that based on inverse and predictive modeling results, the SOM successfully depicted arsenite BTCs from several soil columns and is thus recommended for describing arsenite transport in soils.

In the second example, Elbana and Selim (2012) carried out miscible displacement column experiments and batch adsorption to assess Cu mobility and reactivity in calcareous soils. A second objective was to examine the

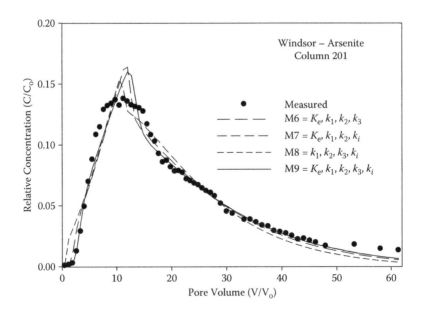

FIGURE 6.23
Comparison of second-order model simulations using several model versions for describing the arsenite breakthrough curve (BTC) from Windsor soil.

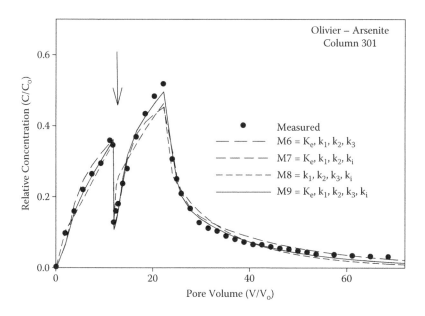

FIGURE 6.24

Comparison of second-order model simulations using several model versions for describing the arsenite breakthrough curve (BTC) from Olivier soil. The arrow indicates pore volumes when flow interruptions occurred.

prediction capability of SOM in describing Cu mobility in strongly reactive calcareous soils. The strongest Cu retention was observed in the surface soil layer having 2.76% $CaCO_3$ in comparison to the subsurface layer having 1.18% $CaCO_3$. Based on the Cu BTC results shown in Figure 6.25, recovery in the effluent was only 27% of that applied. Such low Cu recovery in the effluent was not surprising for this calcareous surface soil. In contrast Cu recovery was 60% of that applied for the subsurface soil. Rodriguez-Rubio et al. (2003) suggested that Cu was preferentially retained in calcareous soils through precipitation of CuO, $Cu_2(OH)_2CO_3$, or $Cu(OH)_2$ and by adsorption on soil carbonates. Elzinga and Reeder (2002) used extended x-ray absorption fine-structure (EXAFS) spectroscopy to characterize Cu adsorption complexes at the calcite surface. They observed that Cu occupied Ca sites in the calcite structure, and formed inner-sphere Cu adsorption complexes at calcite surfaces. EXAFS results revealed that the precipitation of malachite ($Cu_2(OH)_2CO_3$) did not take place in Cu/calcite suspensions at Cu concentration of 5.0 µM and 10.0 µM.

The solid curves in Figure 6.25 are simulations using the SOM model. Comparison of calculated BTCs and measured Cu effluent concentrations illustrates the capability of the SOM model in predicting Cu mobility in the surface and subsurface columns. Simulations using a linear model are also given in Figure 6.25, indicated by the dashed curves. The linear model

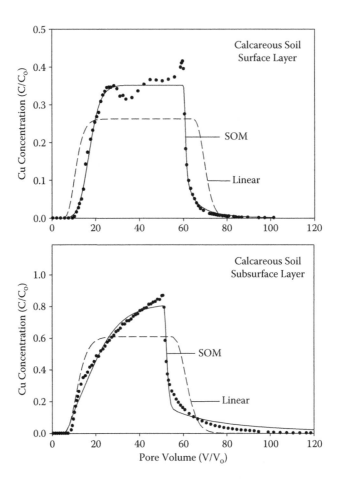

FIGURE 6.25
Breakthrough results of Cu from surface soil (top) and subsurface soil (bottom). Solid curves are simulations using the second-order model. Dashed curves represent linear model predictions.

used by Elbana and Selim (2012) is also described by Toride, Leij, and van Genuchten (1995). This analytical model was utilized to solve the inverse problem based on the assumption of linear equilibrium sorption and an irreversible reaction to account for strong Cu sorption. The linear model used, which incorporates a first-order decay (sink) term, is

$$R\frac{\partial C}{\partial t} = D\frac{\partial^2 C}{\partial z^2} - v\frac{\partial C}{\partial z} - \mu C \qquad (6.43)$$

where R is a dimensionless retardation factor ($R = 1 + \rho K_d/\phi$), and K_d is a partitioning coefficient (mL g^{-1}). In Equation 6.44 the term R accounts for linear

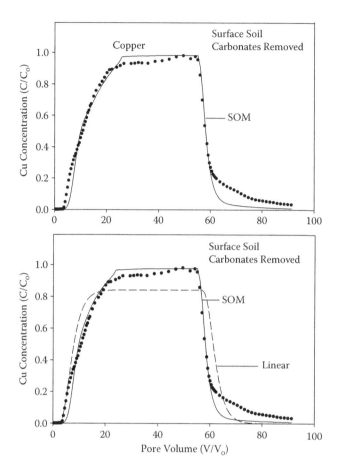

FIGURE 6.26
Breakthrough results of Cu and Ca from the surface soil after removed of carbonate. Solid curves are simulations using the second-order model. The dashed curve represents linear model predictions.

equilibrium sorption. The rate coefficient μ (h^{-1}) associated with the sink term (μC) captures irreversible retention (or removal) of a chemical directly from the soil solution based on first-order kinetics. The simulations shown indicate early arrival of the Cu BTCs and failure to describe the tailing of the leaching right side of the BTCs. In addition, concentration peak maxima of the BTCs were underestimated.

Elbana and Selim (2012) investigated the influence of the removal of carbonates from the surface soil on Cu mobility, illustrated by the BTC shown in Figure 6.26. The BTC is characterized by early arrival of Cu in the effluent (three pore volumes) and peak concentration C/C_o of 1.0. Moreover, 87% of the applied Cu was recovered in the effluent solution. In contrast only

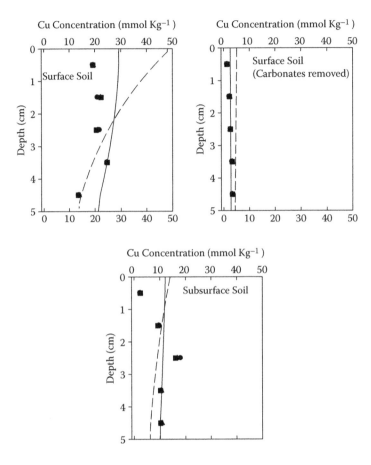

FIGURE 6.27
Copper sorbed versus column depth based on soil extractions. Solid and dashed curves represent second-order and linear model predictions, respectively. Symbols are experimental measurements and represent different replications.

27% was recovered when carbonates were not removed. Good description of the BTC was obtained using the SOM. In addition, use of the linear model provided good overall description of the BTC although it failed to describe the slow release of Cu during leaching. The capability of SOM in describing the distribution of Cu retained versus soil depth following the termination of the miscible displacement experiments is illustrated in Figure 6.27. The amount of Cu retained ranged from 13.60 to 24.65 mmol kg^{-1}. As expected, when the carbonates were removed, a low amount of Cu was retained in the soil (<3.5 mmol kg^{-1}). In general, the amount of Cu retained by each soil is reflected by the amount leached or recovered in the effluent solution. Sorbed Cu versus depth using the SOM and linear models are given by the solid and dashed curves, respectively. Simulated results showed good predictions of Cu distribution for the surface soil when the carbonates were removed. In contrast, poor predictions were realized for the surface when carbonate was not removed.

References

Amacher, M. C., and H. M. Selim. 1994. Mathematical models to evaluate retention and transport of chromium(VI) in soils. *J. Ecol. Modeling*. 74: 205–230.

Arai, Y., and D. L. Sparks. 2002. Residence time effect on arsenate surface precipitation at the aluminum oxide-water interface. *Soil Sci*. 167: 303–314.

Brenner, H. 1962. The diffusion model of longitudinal mixing in beds of finite length: Numerical values. *Chem. Eng. Sci*. 17: 220–243.

Elbana, T. A., and H. M. Selim. 2012. Copper transport in calcareous soils: Miscible displacement experiments and second-order modeling. *Vadose Zone J*. 11: 212–231.

Elkhatib, E. A., O. L. Bennett, and R. J. Wright. 1984. Kinetics of arsenite adsorption in soils. *Soil Sci. Soc. Am. J*. 48: 758–762.

ElPrince, A. M., and G. Sposito. 1981. Thermodynamic derivation of equations of the Langmuir type for ion exchange equilibria in soils. *Soil Sci. Soc. Am. J*. 45: 277–282.

Elzinga, E. J., and R. J., Reeder. 2002. X-ray absorption spectroscopy study of Cu^{2+} and Zn^{2+} adsorption complexes at the calcite surface: Implications for site specific metal incorporation preferences during calcite crystal growth. *Geochim. Cosmochim. Acta* 66: 3943–3954.

Hinz, C., B. Buchter, and H. M. Selim. 1992. Heavy metal retention in soils: Application of multisite models to zinc sorption. In *Engineering Aspects of Metal-Waste Management*, edited by I. K. Iskandar and H. M. Selim, 141–170. Boca Raton, FL: Lewis Publishers.

Lapidus, L., and N. R. Amundson. 1952. Mathematics of adsorption in beds. VI. The effect of longitudinal diffusion in ion exchange and chromatographic column. *J. Phys. Chem*. 56: 84–988.

Nielsen, D. R., M. Th. van Genuchten, and J. M. Biggar. 1986. Water flow and transport processes in the unsaturated zone. *Water Resour. Res*. 22: 895–1085.

Ma, L., and H. M. Selim. 1994. Predicting atrazine adsorption-desorption in soils: A modified second order model. *Water Resour. Res*. 30: 447–456.

Ma, L., and H. M. Selim. 1998. Coupling of retention approaches to physical nonequilibrium models. In *Physical Nonequilibrium in Soils: Modeling and Application*, edited by H. M. Selim and L. Ma, 83–115. Chelsea, MI: Ann Arbor Press.

Murali, V., and L. A. G. Aylmore. 1983. Competitive adsorption during solute transport in soils. I. Mathematical models. *Soil Sci*. 135: 143–150.

Raven, K. P., A. Jain, and R. H. Loeppert. 1998. Arsenite and arsenate adsorption on ferrihydrite: Kinetics, equilibrium, and adsorption envelopes. *Environ. Sci. Technol*. 32: 344–349.

Rodriguez-Rubio, P., E. Morillo, L. Madrid, T. Undabeytia, and C. Maqueda. 2003. Retention of copper by a calcareous soil and its textural fractions: Influence of amendment with two agroindustrial residues. *Eur. J. Soil Sci*. 54: 401–409.

Rubin, J. 1983. Transport of reactive solutes in porous media, relation between mathematical nature of problem formulation and chemical nature of reactions. *Water Resour. Res.* 19: 1231–1252.

Selim, H. M., and M. C. Amacher. 1988. A second-order kinetic approach for modeling solute retention and transport in soils. *Water Resour. Res.* 24: 2061–2075.

Selim, H. M., and M. C. Amacher. 1997. *Reactivity and Transport of Heavy Metals in Soils.* Boca Raton, FL: CRC Press.

Selim, H. M., M. C. Amacher, and I. K. Iskandar. 1990. *Modeling the Transport of Heavy Metals in Soils.* U. S. Army Corps of Engineers, Cold Regions Research & Engineering Laboratory, Monograph 90-2.

Selim, H. M., J. M. Davidson, and R. S. Mansell. 1976. Evaluation of a 2-site adsorption-desorption model for describing solute transport in soils. In Proc. Summer Computer Simulation Conf., Washington, D.C. July 12–14, 1976, 444–448. La Jolla, CA, Simulation Councils Inc.

Selim, H. M., and L. Ma. 2001. Modeling nonlinear kinetic behavior of copper adsorption-desorption in soil. In *Physical and Chemical Processes of Water and Solute Transport/Retention in Soil.* SSSA special publication no. 56, edited by H. M. Selim and D.L. Sparks, 189–212. Madison, WI: Soil Science Society of America.

Selim, H. M., and R. S. Mansell. 1976. Analytical solution of the equation of reactive solutes through soils. *Water Resour. Res.* 12: 528–532.

Selim, H. M., and H. Zhang. 2007. Arsenic adsorption in soils: Second-order and multireaction models. *Soil Sci.* 72: 444–458.

Toride, N., F. J. Leij, and M. T. van Genuchten. 1995. The CXTFIT code for estimating transport parameters from laboratory or field tracer experiments version 2.0. U.S. Salinity Lab., U.S. Department of Agriculture, ARS, Riverside, CA.

Zhang, H., and H. M. Selim. 2011. Second-order modeling of arsenite transport in soils. *J. Contaminant Hydrol.* 126: 121–129.

7

Competitive Sorption and Transport

Over the last three decades, several studies were carried out to identify the dominant mechanisms controlling the fate and overall behavior of heavy metals in the soil-water environment. Most investigations focused on describing the sorption and transport of heavy metals under field conditions and in laboratory and greenhouse experiments. Adsorption-desorption on the surface of solid minerals and organic matters is one of the dominant reactions impacting the fate and transport of heavy metals in soils. Most efforts focused on describing the transport and retention of one heavy metal species only. Such an assumption implies that all other interactions that occur in the soil do not greatly influence the behavior of the heavy metal species under consideration. This simplification is unrealistic and does not represent the soil environment, which contains many chemical species having various interactions.

It is generally accepted that competing ions strongly affect heavy metal retention and release in soils. Industrial waste and sewage sludge disposed of on land often contain appreciable amounts of heavy metal such as Cu, Zn, Cd, and Ni and thus create a risk for croplands, as well as animals and humans. In most cases, soil contamination involves several heavy metals, that is, a multiple-component system. In fact, competition among heavy metal species present in the soil solution for available adsorption sites on soil matrix surfaces is a commonly observed phenomenon (Murali and Aylmore, 1983; Kretzschmar and Voegelin, 2001, among others). Competitive adsorption and desorption processes of heavy metals in minerals and soil organic matter have significant effects on their fate and mobility in soils and aquifers. The selective sorption among competing heavy metal ions may greatly impact their bioavailability in soils (Gomes, Fontes, and da Silva, 2001). Enhanced mobility as a result of competitive sorption has been often observed for several trace metal contaminants in soils.

The adsorption of heavy metal by clay minerals, metal oxides, and organic materials has generally been explained with two types of reaction mechanisms: (1) ion exchange in the diffuse layer as a result of electrostatic force and (2) surface complexation through the formation of strong covalent bonds between heavy metal ions and specific reaction sites on the surfaces of minerals or organic matters. The ion exchange reaction is also referred to as nonspecific sorption and the surface complexation is referred to as specific sorption (Sposito, 1994; Sparks, 1998). A variety of models have been developed with the specific goal of predicting competitive sorption behavior of heavy metals in soils. Equilibrium ion exchange models and surface complexation models were incorporated into geochemical models to simulate the reactions among multiple heavy metals in the soil-water environment (Allison, Brown, and Novo-Gradac, 1991; Parkhurst and Appelo, 1999). Such models were used to simulate, with varying degrees of success, the competitive sorption and transport of several heavy metals in soils and aquifers (e.g., Smith and Jaffe, 1998; Serrano et al., 2005). However, geochemically based models require a detailed description of the chemical and mineral composition of solution and porous media, as well as numerous reaction constants. In fact, several of the required parameters are either unavailable or unreliable under most conditions (Goldberg and Criscenti, 2008). Moreover, the inherent heterogeneity of natural porous media often impedes the application of chemical reaction based models. As a result, sorption reactions are frequently simulated using empirical models of the equilibrium type, such as the Freundlich and Langmuir.

In contrast to empirical models, in most geochemical models, sorption processes are often considered as instantaneous where equilibrium conditions are attained in a relatively short time (minutes or several hours). Traditionally, heavy metal sorption studies have been carried out based on batch equilibration experiments within a short period of reaction time (hours or days). In recent years, numerous studies have demonstrated the lack of reaction equilibrium of contaminants in soils (Selim, 1992). Laboratory experiments using kinetic approaches have demonstrated that sorption of most reactive chemicals in soils was time dependent at various time scales (Selim and Amacher, 1997). For several heavy metals, a slow but significant reaction phase may exist due to (1) the transport of ion species from bulk solution to the reaction sites on mineral surfaces and (2) chemical kinetics of reactions such as ion exchange, formation of inner-sphere surface complexes, precipitation into distinct solid phases, or surface precipitation on minerals. According to Sparks (1998), the kinetics of adsorption-desorption need to be considered for accurate simulation of the fate of heavy metals in the soil-water environment.

7.1 Transport Equation

Competitive sorption of interacting ions often results in complex breakthrough patterns during their transport in soils and geological media. Therefore, describing heavy metal transport requires retention models that account for their governing mechanisms. The one-dimensional reactive-convective-dispersive transport equation is the most frequently used model for describing the transport of dissolved chemicals in soils (Selim, 1992).

$$\frac{\partial \theta C_i}{\partial t} + \rho \frac{\partial S_i}{\partial t} = \frac{\partial}{\partial z} \theta D \frac{\partial C_i}{\partial z} - \frac{\partial q C_i}{\partial z} \tag{7.1}$$

where S_i is the amount of adsorption (mg kg^{-1}), C_i is the dissolved concentration (mg L^{-1}), i indicates the ith component in the system, D is the dispersion coefficient (cm^2 h^{-1}), θ is the soil moisture content (cm^3 cm^{-3}), ρ is the soil bulk density (g cm^{-3}), z is distance (cm), and t is reaction time (h). Retention reactions of a solute from the soil solution by the matrix of soils and geological media is accounted for by the term ($\frac{\partial S_i}{\partial t}$) in Equation 7.1 and can be quantified based on several approaches. A number of transport models simulate heavy metal sorption based on the local equilibrium assumption (LEA). Here one assumes that the reaction of an individual solute species in the soil is sufficiently fast and that an apparent equilibrium may be observed in a time scale considerably shorter than that of the transport processes. The local equilibrium assumption is the basis for several commonly used models, including ion-exchange, surface complexation, Freundlich, and Langmuir models. A discussion of the various models from the perspective of competitive sorption and transport is given in subsequent sections. In contrast to the LEA, for most heavy metals time-dependent retention in soils has been commonly observed, as discussed in previous chapters. As a result, a number of formulations were introduced to describe their kinetic sorption behavior in soils. Examples of kinetic models include the first-order, Freundlich kinetic, irreversible, and second-order models. Commonly used equilibrium and kinetic models are summarized in Table 7.1.

Another class of models is that of the multireaction, multistep, or multisite equilibrium-kinetic models. Basic to multireaction and multisite models is that the soil is a heterogeneous system with different constituents (clay minerals, organic matter, Fe and Al oxides, carbonates, etc.), with sites having different affinities/energies for heavy metal sorption. Therefore, a heavy metal species is likely to react with various constituents (sites) by different mechanisms. As a result, a single equilibrium or chemical kinetic reaction is unlikely to be the dominant mechanism in heterogeneous soils, which are made of multiple sites having different energies for heavy metal sorption. In subsequent sections, multireaction and multisite models and simulations in competitive systems are presented.

TABLE 7.1

Competitive Equilibrium and Kinetic Retention Models

Model	Formulation
Sheindorf-Rebhun-Sheintuch (SRS) equilibrium model	$S_i = K_i C_i \left(\sum\limits_{j=1}^{I} \alpha_{i,j} C_j \right)^{n_i - 1}$
Sheindorf-Rebhun-Sheintuch (SRS) kinetic model	$\dfrac{\partial (S_1)_i}{\partial t} = k_{1,i} \dfrac{\theta}{\rho} C_i \left(\sum\limits_{j=1}^{I} \alpha_{i,j} C_j \right)^{n_i - 1} - k_{2,i} (S_1)_i$
Competitive Langmuir equilibium model	$\dfrac{S_i}{S_{\max}} = \dfrac{K_i C_i}{1 + \sum_j K_j C_j}$
Competitive Langmuir kinetic model	$\dfrac{\partial S_i}{\partial t} = (\lambda_f)_i \dfrac{\theta}{\rho} C_i \left(S_{\max} - \sum\limits_{i=1}^{I} S_j \right) - (\lambda_b)_i S_i$
Vanselow ion exchange	${}^v K_{ij} = \left(\dfrac{(a_i^*)^{vj}}{(a_j^*)^{vi}} \right) \left(\dfrac{(\zeta_j)^{vj}}{(\zeta_j)^{vi}} \right)$
Gaines and Thomas ion exchange	${}^G K_{ij} = \dfrac{(\gamma_j)^{vi}}{(\gamma_i)^{vj}} \left(\dfrac{s_i}{C_i} \right)^{vj} \left(\dfrac{s_j}{C_j} \right)^{vi}$
Rothmund-Kornfeld ion exchange	$\dfrac{(s_i)^{vj}}{(s_j)^{vi}} = {}^R K_{ij} \left[\dfrac{(c_i)^{vj}}{(c_j)^{vi}} \right]^{n}$
Elovich ion exchange	$\dfrac{\partial S_i}{\partial t} = a\, e^{-BS_i}$
Factional power model	$S_i = \kappa t^{\beta}$
Parabolic diffusion model	$\dfrac{S_i}{(S_{eq})_i} = \dfrac{4}{\sqrt{\pi}} \sqrt{\dfrac{D_m t}{r^2}} - \dfrac{D_m t}{r^2}$

7.2 Equilibrium Ion Exchange

One of the earliest concepts of competition among the various ions in the soil solution and those retained on matrix surfaces is that of ion exchange. Here one assumes that a positive (or negative) ion in solution can exchange with another ion retained on charged surfaces (Sparks, 2003). Ion exchange is commonly considered an instantaneous process representing (nonspecific) sorption mechanisms and as a fully reversible reaction between heavy metal ions in the soil solution and those retained on charged surfaces of the soil matrix. In a standard mass action formulation, the exchange reaction for two competing ions i and j, having valencies v_i and v_j, respectively, may be written as:

$$^{T}K_{ij} = \frac{\left(a_i^*\right)^{vj}}{\left(a_j^*\right)^{vi}} \tag{7.2}$$

where $^{T}K_{ij}$ denotes the thermodynamic equilibrium constant and α^* (omitting the subscripts) are the ion activity in soil solution and on the exchanger surfaces, respectively. Based on Equation 7.2, one can denote the parameter $^{v}K_{ij}$ as:

$$^{v}K_{ij} = \frac{^{T}K_{ij}}{\frac{(\zeta_j)^{vj}}{(\zeta_j)^{vi}}} \tag{7.3}$$

where $^{v}K_{ij}$ is the Vanselow selectivity coefficient and ζ the activity coefficient on the soil surface. It is recognized that in soils, ion exchange involves a wide range of thermodynamically different sites. As a result, a common practice is to ignore the activity coefficients of the adsorbed phase (ζ) in general. In addition, the much simpler Gaines and Thomas (1953) selectivity coefficient $^{G}K_{ij}$ may be used, where:

$$^{G}K_{ij} = \frac{(\gamma_j)^{vi}}{(\gamma_i)^{vj}} \left(\frac{s_i}{C_i}\right)^{vj} \left(\frac{s_j}{C_j}\right)^{vi} \tag{7.4}$$

This formulation is more conveniently incorporated into the dispersion-convection transport equation (7.1). In Equation 7.4, γ_i and γ_j are dimensionless solution-phase activity coefficients where $a_i = \gamma_i C_i$. In addition, the terms s_i and s_j are dimensionless, representing the solid-phase concentrations expressed in terms of equivalent fraction, $s_i = S_i / \Omega$. Here the term Ω is the cation exchange (or adsorption) capacity of the soil (mmol$_c$ kg^{-1} soil) and S_i is the concentration (mmol$_c$ kg^{-1}) of adsorbed-phase soil. Although Ω is often assumed as invariant, it is recognized that Ω has been observed to be dependent on soil pH and the counterions present in the soil. Moreover, there are several other ways to express the adsorbed-phase concentration on a fractional basis, including as a molar rather than as an equivalent value.

For the simple case of binary homovalent ions, that is, $v_i = v_j = v$, and assuming similar ion activities in the solution phase ($\gamma_i = \gamma_j = 1$), Equation 7.3 can be rewritten as:

$$K_{ij} = \left(\frac{s_i}{C_i}\right) \bigg/ \left(\frac{s_j}{C_j}\right) \tag{7.5}$$

where K_{ij} represents the affinity of ion i over j or a separation factor for the affinity of ions on exchange sites. Rearrangement of Equation 7.5 yields the following isotherm relation for equivalent fraction of ion 1 as a function of c_1:

$$s_1 = \frac{K_{12}\,c_1}{1+(K_{12}-1)\,c_1} \tag{7.6}$$

and c_1 relative concentration (dimensionless) and C_T (mmol$_c$ L^{-1}) represent the total solution concentration:

$$c_i = \frac{C_i}{C_T} \quad and \quad C_T = \sum_i C_i \tag{7.7}$$

The respective isotherm equation for ion 2 (i.e., s_2 versus c_2) can be easily deduced. This dimensionless isotherm relation is represented in Figure 7.1 for different values of K_{12}. For $K_{12} = 1$, a linear isotherm relation is produced, represented by the solid line in Figure 7.1. This clearly illustrates a 1:1 relationship between relative concentration in solution and that on the adsorbed phase (i.e., $s_1 = c_1$). This also implies that the two ions 1 and 2 each have equal affinity for the exchange sites. In contrast, for K_{12} different than 1, we have nonlinear sorption isotherms. Specifically, for $K_{12} > 1$, sorption of ion 1 is preferred and the isotherms are convex. For $K_{12} < 1$, sorption affinity is apposite and the isotherms are concave. Examples of homovalent ion exchange isotherms are illustrated in Figure 7.2 for Ca-Mg in soils and clay mineral

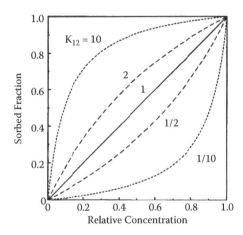

FIGURE 7.1
Exchange isotherms as affected by different values of selectivity coefficients (K_{12}). (From H. M. Selim, and M. C. Amacher. 1997. *Reactivity and Transport of Heavy Metals in Soils*. Boca Raton, FL: CRC Press. With permission.)

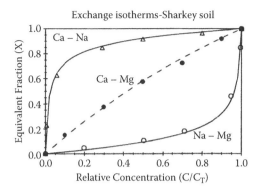

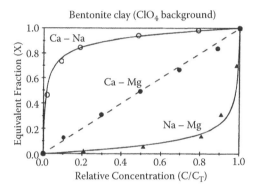

FIGURE 7.2
Exchange isotherms for Mg-Na, Ca-Mg, and Na-Ca on Sharkey soil (A) and bentonite-sand mixture (B). Smooth curves are predictions obtained using the constant exchange selectivity model.(From H. M. Selim and M. C. Amacher. 1997. *Reactivity and Transport of Heavy Metals in Soils*. Boca Raton, FL: CRC Press. With permission.)

systems (Gaston and Selim, 1990a, 1990b) and in Figure 7.3 for Cd-Ca in two soils (Selim et al., 1992).

Examples of predictions of cation breakthrough curves (BTCs) using ion exchange models compared to experimental data for Mg plus Na leached by Ca in a Sharkey clay soil column is shown in Figure 7.4 (Gaston and Selim, 1990a). Breakthrough of Na was early and well described by the model. Moreover, predictions of Ca and Mg results were equally well described. Such predictions are somewhat surprising achievements of a simple transport model based on ion exchange.

Predicting cation transport in soils as shown in Figure 7.4 requires knowledge of several chemical and physical properties of the porous medium, including, at a minimum, the cation-exchange capacity (CEC), exchange selectivity coefficients, bulk density, and the dispersion coefficient. The CEC and the physical properties are readily determined. Estimation of

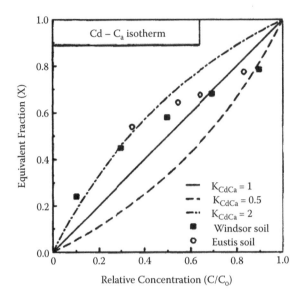

FIGURE 7.3
Cadmium-calcium exchange isotherm for Windsor and Eustis soils. Solid and dashed curves are simulations using different selectivities (K_{CdCa}). (From H. M. Selim and M. C. Amacher. 1997. *Reactivity and Transport of Heavy Metals in Soils.* Boca Raton, FL: CRC Press. With permission.)

cation-exchange selectivities, however, typically requires development of exchange isotherms. The experimental methods are laborious. If exchange selectivities for reference materials could be used to obtain fairly accurate predictions of cation movement in field soils, then it might be feasible to

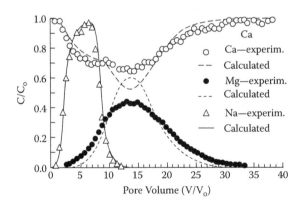

FIGURE 7.4
Breakthrough results for a ternary system (Na, Ca, and Mg) in a Sharkey clay soil. Solid curves are predictions based on the equilibrium ion exchange transport model.

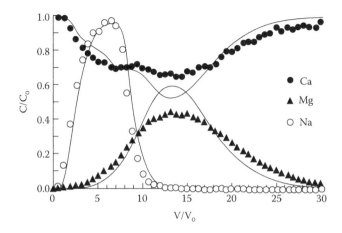

FIGURE 7.5
Breakthrough results for ternary system (Na, Ca and Mg) in a Sharkey clay soil. Predictions are based on ion exchange selectivity coefficients for bentonite clay.

broaden the database of land management decisions for agricultural production or waste disposal to include such predictions. The applicability of selectivity parameters of a common type of mineral for prediction of cation motilities of soils having mineralogies dominated by similar minerals showed success (see Gaston and Selim, 1990b, 1991). For example, transport model predictions based on selectivity coefficients of pure montmorillonite shown in Figure 7.2 well described cation leaching in columns of bulk samples of predominantly montmorillonitic Sharkey soil as shown in Figure 7.5. As reported by Gaston and Selim (1991) good cation predictions were obtained for a predominantly kaolinitic Mahan soil when selectivity coefficients were based on pure kaolinite.

Several studies indicated that the affinity of heavy metals to soil matrix surfaces increases with decreasing heavy metal fraction on exchanger surfaces (Abd-Elfattah and Wada, 1981; Harmsen, 1977; Selim et al., 1992; Hinz and Selim, 1994). Using an empirical selectivity coefficient, it was shown that Zn affinity increased up to two orders of magnitude for low Zn surface coverage in a Ca background solution (Abd-Elfattah and Wada, 1981). Mansell et al. (1988) successfully relaxed the assumption of constant affinities and allowed the selectivity coefficients to vary with the amount adsorbed on the exchange surfaces. The Rothmund-Kornfeld selectivity coefficient incorporates variable selectivity based on the amount of adsorbed or exchanger composition. The approach is empirical and provides a simple equation that incorporates the characteristic shape of binary exchange isotherms as a function of equivalent fraction of the amount sorbed as well as the total solution concentration in solution. Harmsen (1977) and Bond and Phillips (1990) expressed the Rothmund-Kornfeld selectivity coefficient as:

$$\frac{(s_i)^{vj}}{(s_j)^{vi}} = {}^{R}K_{ij}\left[\frac{(c_i)^{vj}}{(c_j)^{vi}}\right]^{n} \tag{7.8}$$

where n is a dimensionless empirical parameter associated with the ion pair i-j and ${}^{R}K_{ij}$ is the Rothmund-Kornfeld selectivity coefficient. The above equation is best known as a simple form of the Freundlich equation, which applies to ion exchange processes. As pointed out by Harmsen (1977), the Freundlich equation may be considered as an approximation of the Rothmund-Kornfeld equation valid for $s_i \ll s_j$ and $c_i \ll c_j$, where:

$$s_i = {}^{R}K_{ij}(c_i)^{n} \tag{7.9}$$

The ion exchange isotherms in Figure 7.6 show the relative amount of Zn and Cd adsorbed as a function of relative solution concentration along with best-fit isotherms based on the Rothmund-Kornfeld equation for two acidic soils (Hinz and Selim, 1994). The diagonal line represents a non-preference isotherm (${}^{R}K_{ij} = 1$, $n = 1$) where competing ions (Ca-Zn or Ca-Cd) have equal affinity for exchange sites. The sigmoidal shapes of the isotherms reveal that Zn and Cd sorption exhibit high affinity at low concentrations, whereas Ca exhibits high affinity at high heavy metal concentrations. This behavior is well described by the Rothmund-Kornfeld isotherm with n less than one. Isotherms in Figure 7.6 are consistent with those of Bittel and Miller (1974), Harmsen (1977), and Abd-Elfattah and Wada (1981). However, Abd-Elfattah and Wada obtained much larger selectivities for low heavy metal concentrations on exchange surfaces compared to the results shown.

7.3 Kinetic Ion Exchange

Numerous studies have indicated that ion exchange is a kinetic process in which equilibrium is not instantaneously reached. Observed kinetic or time-dependent ion exchange behavior is most likely a result of the transport of initially sorbed ions from exchange sites to the bulk solution and the reverse process of the transport of replacing ion from bulk solution to exchange sites. Specifically, the rate of ion exchange is dependent on the following processes: (1) diffusion of ions in the aqueous solution, (2) film diffusion at the solid-liquid interface, (3) intraparticle diffusion in micropores and along pore wall surfaces, and (4) interparticle diffusion inside solid particles (Sparks, 1989). Due to the complexity of the kinetic processes, over the last three decades

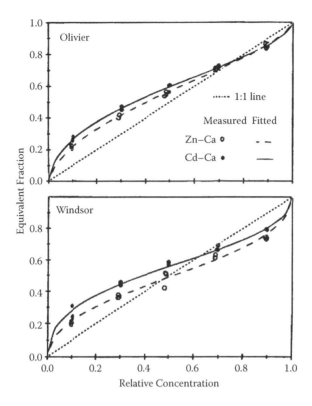

FIGURE 7.6
Ion exchange isotherms of Cd-Ca and Zn-Ca for Olivier and Windsor soils (relative concentration (C/C_T) versus the sorbed fraction (s/Ω). Solid and dashed curves are fitted using the Rothmund-Kornfeld equation. (From H. M. Selim and M. C. Amacher. 1997. *Reactivity and Transport of Heavy Metals in Soils*. Boca Raton, FL: CRC Press. With permission.)

several kinetic rate formulations have been proposed. Such formulations have been applied to describe sorption-desorption kinetic results for heavy metal ions. The pseudo first-order or mass action model is perhaps the most commonly used approach, where one assumes the kinetic rate is a function of the concentration gradient on the exchange surfaces. For the exchangeable amount S (omitting the subscript i) the rate equation can be expressed as (Selim et al. 1992):

$$\frac{\partial S}{\partial t} = \alpha(S_{eq} - S) \tag{7.10}$$

where α is an apparent rate coefficient (d^{-1}) for the kinetic-type sites. For large values of α, S approaches S_{eq} in a relatively short time and equilibrium is rapidly achieved. In contrast, for small α, kinetic behavior should be dominant for extended periods of time. Illustrative examples of utilizing the kinetic ion exchange concept are presented in a later section. It should also be mentioned that expressions similar to the above model have been used to describe mass transfer between mobile and immobile water in porous media (Coats and Smith, 1964), as well as chemical kinetics (Parker and Jardine, 1986).

Other kinetic expressions have also been employed to describe the kinetic exchange of ions in mineral and soils. The parabolic diffusion model is based on the assumption of diffusion controlled rate-limited processes in media with homogeneous particle size. The parabolic diffusion equation was derived from Fick's second law of diffusion in a radial coordinate system.

$$\frac{S}{(S_{eq})_i} = \frac{4}{\sqrt{\pi}} \sqrt{\frac{D_m t}{r^2}} - \frac{D_m t}{r^2} \tag{7.11}$$

where r is the average radius of soil or mineral particle, and D_m is the molecular diffusion constant. The Elovich model is anther empirical kinetic retention model, which may be expressed as:

$$\frac{\partial S_i}{\partial t} = a\, e^{-BS_i} \tag{7.12}$$

where a is the initial adsorption rate and B is an empirical constant. Jardine and Sparks (1984) have compared the first-order, parabolic diffusion, and Elovich approaches described above to describe the kinetic exchange of K-Ca in clay minerals and soils. They found that the pseudo first-order model provided the best overall goodness-of-fit of the experimental results. Recently, a fractional power approach was introduced by Serrano et al. (2005) having the form:

$$S = \kappa t^{\beta} \tag{7.13}$$

where κ and β are empirical constants. They compared the overall sorption kinetics of Pb and Cd for single and binary systems. Their results showed that the simultaneous presence of the competing metal did not affect the estimated apparent sorption rate, which indicated that the rate-limiting processes of the sorption of heavy metal ions were not impacted by the competing ions.

7.4 Examples

Several examples are given here to illustrate the capability of the competitive model to describe heavy metals transport in soils. Figure 7.7 is an example of Cd miscible displacement results for Eustis fine sandy soil from Selim et al. (1992). The sequence of input solutions was 10 mmol$_c$ L^{-1} of Ca(NO$_3$)$_2$ followed

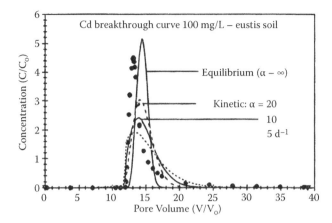

FIGURE 7.7
Measured (closed circles) and predicted breakthrough curves in Windsor soil column. Curves are predictions using equilibrium and kinetic ion exchange with different α values. (From H. M. Selim and M. C. Amacher. 1997. *Reactivity and Transport of Heavy Metals in Soils*. Boca Raton, FL: CRC Press. With permission.)

by 1.786 $mmol_c$ L^{-1} $Cd(NO_3)_2$ then 10 $mmol_c$ L^{-1} of $Ca(NO_3)_2$ subsequently added. As a result of changes in the total concentration (C_T), a pronounced snowplow effect was observed as shown in Figure 7.7. Here the measured peak concentration was about five times the input Cd pulse. Moreover, use of the ion exchange approach with the assumption of local equilibrium adequately predicted the shape and location of the BTC, although the BTC was somewhat more retarded than observed. The selectivity coefficient (K_{12}) used in the simulations shown in Figure 7.5 was equal to 7.1, indicating equal affinity of Cd and Ca to the soil surface. It is clear that the use of the kinetic ion exchange improved overall prediction of the BTC even though this resulted in the lowering of peak concentrations. Therefore, the competitive ion exchange approach was capable of predicting the snowplow elution of the Cd pulse from the soil column.

To further test the capability of the competitive model, two data sets from multiple pulse applications are illustrated. Figures 7.8 and 7.9 are for Windsor soil where Cd pulse applications were 10 and 100 mg L^{-1}, respectively (Selim et al., 1992). For all multiple pulses, the ion exchange approach well predicted the position of the BTC peaks. In fact, the assumption of equilibrium ion exchange adequately predicted the observed snow plow effect for the two Windsor data sets. Good predictions were also obtained for peak maxima for the 100 mg L^{-1} data set (Figures 7.8 and 7.9). Calculated BTCs shown in these figures were obtained from input parameters that were independently determined. Specifically, curve fitting of the data was not implemented to obtain the predictions shown. The shape of the Cd peaks of Figures 7.8 and 7.9 are due in part to the concentration and width or the number of pore

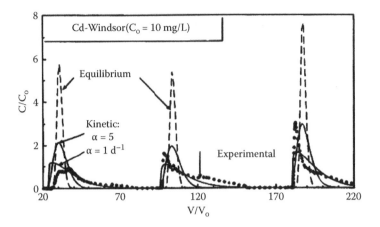

FIGURE 7.8
Measured (closed circles) and predicted breakthrough curves in Windsor soil column for three Cd pulses of $C_0 = 10$ mg L^{-1}. Curves are predictions using equilibrium and kinetic ion exchange with different α values. (From H. M. Selim and M. C. Amacher. 1997. *Reactivity and Transport of Heavy Metals in Soils*. Boca Raton, FL: CRC Press. With permission.)

volumes of each input pulse (Selim et al., 1992). Lower peak concentrations for 10 mg L^{-1} in comparison to 100 mg L^{-1} is perhaps related to the equivalent fraction of Cd on exchange surfaces for Windsor soil. In Figure 7.8, the maximum C/C_0 increased from 0.9 to 1.6 and reached 3.0 for the third Cd

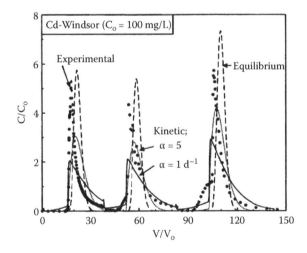

FIGURE 7.9.
Measured (closed circles) and predicted breakthrough curves in Windsor soil column for three Cd pulses of $C_0 = 100$ mg L^{-1}. Curves are predictions using equilibrium and kinetic ion exchange with different α values. (From H. M. Selim and M. C. Amacher. 1997. *Reactivity and Transport of Heavy Metals in Soils*. Boca Raton, FL: CRC Press. With permission.)

pulse. The first pulse (22.6 pore volumes) represents about 4% of the total cation exchange in the soil column, and the second and third pulses (20.8 and 30.3 pore volumes) represent 4% and 6%, respectively. Selim et al. (1992) postulated that applied Cd may occupy specific sorption sites on matrix surfaces. Therefore, irreversible Cd sorption could partly explain the fact that only 80% of applied Cd was recovered in the effluent. For the Windsor BTCs shown in Figure 7.9 (C_o = 100 mg L^{-1}), the first pulse alone was equivalent to 30% of the cation exchange capacity of the soil column and perhaps resulted in higher peak maxima due to Ca-Ca ion exchange.

Examples of transport behavior of Zn when variable ionic strength (or C_T) conditions prevailed in the soil columns are presented in Figures 7.10 and 7.11 for two flow velocities (Hinz and Selim, 1994). Since the total concentration

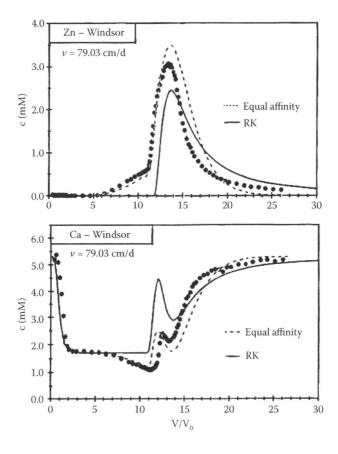

FIGURE 7.10
Zn and Ca breakthrough curves in Windsor soil column at variable ionic strength. Predictions were based on equal affinity (K_1 = 1) and the Rothmund-Kornfeld (RK) equation. (From H. M. Selim and M. C. Amacher. 1997. *Reactivity and Transport of Heavy Metals in Soils*. Boca Raton, FL: CRC Press. With permission.)

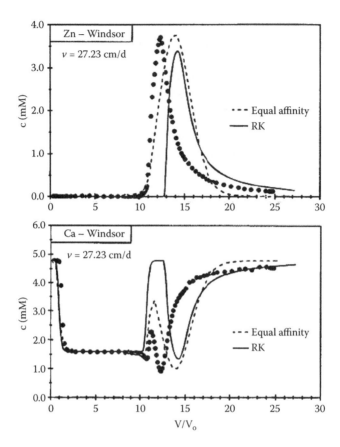

FIGURE 7.11

Zn and Ca breakthrough curves in a Windsor soil column at variable ionic strength. Predictions were based on equal affinity ($K_{12} = 1$) and the Rothmund-Kornfeld approach. (From H. M. Selim and M. C. Amacher. 1997. *Reactivity and Transport of Heavy Metals in Soils.* Boca Raton, FL: CRC Press. With permission.)

of the Zn and Cd input pulse solutions were much lower than that of the displacing Ca solution, chromatographic peaks were observed after 13 pore volumes. The BTCs were somewhat similar in shape except for earlier Zn arrival at high flow velocity. Early appearance of Zn was well described by the predicted BTC (dashed curves) where equal Ca-Zn exchange affinity was assumed. In fact, the chromatographic effect for Ca and Zn was adequately described by the equal affinity BTCs, for both flow velocities. However, the tailing was not well predicted. Contrary to our expectations, the Rothmund-Kornfeld equation predictions (solid curves), for both data sets, were disappointing. The extent of Zn retardation was overestimated by two to three pore volumes (right shift). The opposite trend was observed for Ca, where

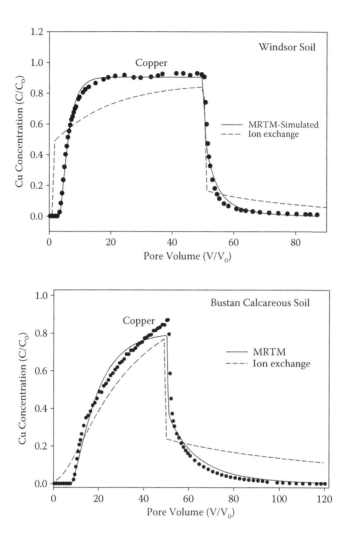

FIGURE 7.12
Breakthrough results of Cu and Ca from Bustan-subsurface calcareous soil and Windsor acidic soil. Multireaction transport model (MRTM) and kinetic ion exchange model simulations are denoted by solid and dashed curves, respectively.

the BTCs were shifted to the left of experimental results. These predictions are indicative of strong Zn affinity (compared to Ca) at low Zn concentrations based on parameter estimates of ion-exchange isotherms using the Rothmund-Kornfeld approach.

Another example that illustrates the application of kinetic ion exchange is that of Elbana and Selim (2011) for Cu transport in different soils (see Figure 7.12). Here Cu solution was introduced in the form of a pulse into columns packed with different calcareous (Bustan) soils and a Windsor soil. Although accurate prediction for Cu transport was not attained for both soils, the assumption that Cu retention occurs via kinetic ion exchange cannot be ignored. The results presented in Figure 7.10 also illustrate the capability of the multireaction and transport model (MRTM) in predicting Cu transport in different soils.

7.5 Competitive Freundlich Model

This model was originally developed by Sheindorf, Rebhun, and Sheintuch (1981) and is commonly referred to as the Sheindorf-Rebhun-Sheintuch (SRS) model. Here competitive or multicomponent sorption is assumed to follow that for single-component sorption based on the Freundlich equation:

$$S = K_F C^b \tag{7.14}$$

The derivation of the SRS equation was based on the assumption of an exponential distribution of adsorption energies for each component. Specifically, the SRS model was developed to describe competitive equilibrium sorption for multicomponent systems where the sorption isotherms of single components follow the Freundlich equation. A general form of the SRS equation can be written as:

$$S_i = K_i C_i \left(\sum_{j=1}^{l} \alpha_{i,j} C_j \right)^{n_i - 1} \tag{7.15}$$

where the subscripts i and j denote metal components i and j, l is the total number of components, and $\alpha_{i,j}$ is a dimensionless competition coefficient for the adsorption of component i in the presence of component j. The parameters K_i and n_i are the Freundlich parameters representing a single component system i as described in Equation 7.14. By definition, $\alpha_{i,j}$ equals 1 when $i = j$. If there is no competition, that is, $\alpha_{i,j} = 0$ for all $j \neq i$, Equation 7.15 yields a single-species Freundlich equation for component i identical to Equation 7.14.

The suitability of the multicomponent SRS equation for describing the competitive adsorption isotherms of trace elements on soil and soil minerals has been investigated by several researchers. A general procedure for applying the SRS equation is first to obtain the Freundlich distribution coefficient K_F and reaction exponent b or n by fitting the single-component isotherms to the

Freundlich equation, followed by estimating the competition coefficients $\alpha_{i,j}$ through fitting the experimental isotherms of binary and ternary mixtures to the SRS equation (Roy, Hassett, and Griffin, 1986a). The competition coefficients $\alpha_{i,j}$ in the SRS equation represent the relative selectivity of the sorbent to the heavy metal species. It is demonstrated that the SRS equation with competition coefficients estimated through nonlinear least squares optimization successfully describes the experimental competitive adsorption isotherms of Ni and Cd on three different soils (Liao and Selim, 2010). Gutierrez and Fuentes (1993) employed the SRS equation to represent the competitive adsorption of Sr, Cs, and Co in Ca-montmorillonite suspensions. They found that the SRS competition coefficients $\alpha_{i,j}$ obtained from experimental data on binary mixtures successfully predicted the competitive adsorption of the ternary mixture Sr-Cs-Co. Similarly, Bibak (1997) found that values of the SRS competitive coefficients obtained from binary sorption experiments successfully predicted sorption data of the ternary solute mixture Cu-Ni-Zn. The SRS equation was successfully used to describe competitive sorption of Cd, Ni, and Zn on a clay soil by Antoniadis and Tsadilas (2007). In addition, the SRS equation was also used by Wu et al. (2002) in representing the competitive adsorption of molybdate, sulfate, selenate, and selenite on γ-Al_2O_3 surface where relative affinity coefficients were used instead of competitive coefficients. The relative affinity coefficients were calculated as the ratios of the proton coefficients of competing anions. The simulation result showed that the sorption affinity of anions on γ-Al_2O_3 surface decreased in the order of $MoO_4^{2-} > SeO_3^{2-} > SeO_4^{2-} > SO_4^{2-}$.

7.6 Competitive Multireaction Model (CMRM)

The kinetic (time-dependent) sorption models of the Freundlich type given were developed to simulate the sorption of solutes during their transport in soils and aquifers (Selim, 1992). Zhang and Selim (2007) extended the equilibrium Freundlich approach to account for kinetic competitive or multiple-component systems. Specifically, the model accounts for equilibrium and kinetic adsorption in a way similar to the SRS equation described above. This model represents a modification of the multireaction and transport model (MRTM), which accounts for equilibrium and kinetic retention of the reversible and irreversible type as discussed in Chapter 5. MRTM accounts for linear as well as nonlinear reaction processes of equilibrium and/or kinetic (reversible and irreversible) type. The model version discussed here is that with reversible as well as irreversible sorption of the concurrent and consecutive type. Here S_e represents the amount retained on equilibrium sites (mg kg^{-1}); S_1 and S_2 represent the amount retained on reversible kinetic sites (mg kg^{-1}); S_s represents the amount retained on consecutive irreversible sites (mg kg^{-1}); and S_{irr} represents the amount retained on concurrent irreversible sites (mg kg^{-1}). The retention reactions associated with the MRTM are similar to those presented earlier in Chapter 3:

$$S_e = K_e \left(\frac{\theta}{\rho} \right) C^b$$

(7.16)

$$\frac{\partial S_1}{\partial t} = k_1 \left(\frac{\theta}{\rho} \right) C^n - k_2 S_1$$

(7.17)

$$\frac{\partial S_2}{\partial t} = \left[k_3 \left(\frac{\theta}{\rho} \right) C^m - k_4 S_2 \right] - k_s S_2$$

(7.18)

$$\frac{\partial S_s}{\partial t} = k_s S_2$$

(7.19)

$$S_{irr} = k_{irr} \left(\frac{\theta}{\rho} \right) C$$

(7.20)

where b is the reaction order; K_e is a dimensionless equilibrium constant (cm^3 g^{-1}); k_1 and k_2 (h^{-1}) are the forward and backward rate coefficients associated with S_1, respectively, and n is the reaction order; k_3 and k_4 (h^{-1}) are the forward and backward rate coefficients associated with S_2, respectively, and m is the reaction order; k_s (h^{-1}) is the irreversible rate coefficient; and k_{irr} (h^{-1}) is the rate coefficient associated with S_{irr}. In the above equations, we assumed $n = b$ since there is no method for estimating n and/or m independently. Zhang and Selim (2007) modified the MRM to account for competitive Freundlich-type retention such that:

$$(S_e)_i = K_{e,i} C_i \left(\sum_{j=1}^{l} \alpha_{i,j} C_j \right)^{n_i - 1}$$

(7.21)

$$\frac{\partial (S_1)_i}{\partial t} = k_{1,i} \frac{\theta}{\rho} C_i \left(\sum_{j=1}^{l} \alpha_{i,j} C_j \right)^{n_i - 1} - k_{2,i} (S_1)_i$$

(7.22)

$$\frac{\partial (S_2)_i}{\partial t} = k_{3,i} \frac{\theta}{\rho} C_i \left(\sum_{j=1}^{l} \alpha_{i,j} C_j \right)^{n_i - 1} - (k_{4,i} + k_{s,i})(S_2)_i \qquad (7.23)$$

$$\frac{\partial (S_s)_i}{\partial t} = k_{s,i} (S_s)_i \qquad (7.24)$$

$$\frac{\partial (S_{irr})_i}{\partial t} = k_{irr,i} \frac{\theta}{\rho} C_i \qquad (7.25)$$

When competition is ignored, that is, $\alpha_{i,j}$ for all $j \neq i$, Equation 71 holds for single-species nth-order kinetic sorption. Examples of the capability of this approach to describe the transport of the competitive arsenate and phosphate behavior in soil columns are presented in a subsequent section.

7.7 Competitive Langmuir Model

The Langmuir equation can be extended to account for competitive sorption of multiple heavy metals in multicomponent systems. In the multicomponent Langmuir approach, one assumes that there is only one set of sorption sites for all competing ions. Furthermore, the model also assumes that the presence of competing ions does not affect the sorption affinity of other ions. Because of these overly simplified assumptions, the modeling ability of the model is rather limited. It should be noted that with the assumption of a fixed number of reaction sites, the surface complexation model described in this chapter gives Langmuir-type adsorption isotherms under constant pH and ionic strength. For time-dependent sorption of competing ions, the multicomponent second-order kinetic equation was proposed in the form of:

$$\frac{\partial S_i}{\partial t} = (\lambda_f)_i \frac{\theta}{\rho} C_i \left(S_{max} - \sum_{i=1}^{l} S_i \right) - (\lambda_b)_i S_i \qquad (7.26)$$

Under equilibrium conditions, Equation 7.26 yields:

$$(K_L)_i = (\lambda_f)_i \frac{(\lambda_f)_i \theta}{(\lambda_b)_i \rho} \qquad (7.27)$$

where λ_f and λ_b (h^{-1}) are the forward and backward rate coefficients, respectively.

Moreover, for the case where one assumes negligible competition among the various ions, Equation 7.26, omitting the subscript i, is reduced to the second-order formulation discussed earlier.

7.8 Experimental Evidence

7.8.1 Nickel and Cadmium

Several studies on Ni and Cd indicate that they exhibit somewhat similar sorption behavior on minerals and soils. These two cations have lower affinities for soil colloids and are generally considered as weakly bonded metals (Atanassova, 1999). A consequence of weakly bonded heavy metals ions such as Cd and Ni is that ion competition may result in their enhanced mobility in the soil environment. Moreover, a number of studies reported varying Cd and Ni affinities in soils and minerals. Several studies indicated that for some soils Cd is of higher affinity than Ni (Gomes et al., 2001; Echeverría et al., 1998; Papini, Saurini, and Bianchi, 2004). Moreover, cation exchange was considered as the major sorption mechanism for both ions. Echeverría et al. (1998) and Antoniadis and Tsadilas (2007) reported that Ni adsorption was stronger than Cd and was related to hydrolysis of divalent ions capable of forming inner-sphere complexes with clay lattice edges. Other studies with minerals, such as kaonilite, montmorillonite, and goethite, indicated stronger affinity for Cd than Ni (Barrow, Gerth, and Brummer, 1989; Puls and Bohn, 1988). For hematite, kinetic sorption results indicated that Ni is of stronger affinity than Cd (Jeon, Dempsey, and Burgos, 2003). Schulthess and Huang (1990) showed that Ni adsorption by clays is strongly influenced by pH as well as silicon and aluminum oxide surface ratios. Moreover, in recent studies using X-ray absorption fine structure (XAFS) and high-resolution transmission electron microscopy (HRTEM) techniques, Ni-Al layered double hydroxide (LDH) was considered responsible for the sorption behavior for pH above 6.5 on pyrophyllite and kaolinite surfaces (Scheidegger, Lamble, and Sparks, 1996; Eick, Naprstek, and Brady, 2001). They suggested that Al dissolved at high pH values could be responsible for Ni precipitate on clay surfaces. The formation of surface-induced precipitates may play an important role in the immobilization of Ni in nonacidic soils. However, surface-induced precipitates were not found for Cd in nonacid soils. This suggests that competitive behavior of Cd and Ni in neutral and alkaline soils may have different characteristics from their behavior in acidic soils.

Modeling competitive adsorption between Ni and Cd in soils has been at best sparse. Nevertheless a few scientists have utilized variable charge surface models and surface complexation models. Barrow, Gerth, and Brummer (1989) successfully utilized a variable-charge surface model in an effort to describe Ni, Zn, and Cd adsorption in a goethite-silicate system. A modified competitive surface complexation model developed by Papini, Saurini, and Bianchi (2004) was adopted to describe competitive adsorption of Pb, Cu, Cd, and Ni by an Italian red soil. Equilibrium and kinetic ion exchange type models were employed by several investigators to describe sorption of heavy metals in soils (Abd-Elfattah and Wada, 1981; Hinz and Selim, 1994). Here the affinity of heavy metals increases with decreasing heavy metal fraction on exchanger surfaces. Using an empirical selectivity coefficient it was shown that Zn affinity increased up two orders of magnitude for low Zn surface coverage in a Ca-background solution (Abd-Elfattah and Wada, 1981). The Rothmund-Kornefeld approach incorporates variable selectivity based on the amount of metal sorbed. Based on the Rothmund-Kornefeld approach, Hinz and Selim (1994) showed strong Zn and Cd affinities at low concentrations.

Another type of competitive adsorption modeling is that based on the Freundlich approach. The Shenindrof-Rebhun-Sheituch (SRS) equation was developed to describe competitive or multicomponent sorption where it was assumed that the single-component sorption follows the Freundlich equation (Sheindorf, Rebhun, and Sheituch, 1981). The derivation of the SRS equation was based on the assumption of an exponential distribution of adsorption energies for each component. Gutierrez and Fuentes (1993) concluded that the SRS approach was suitable in representing competitive adsorption of Sr, Cs, and Co in a system comprising Ca-montmorillonite suspensions. Recently, Antoniadis and Tsadilas (2007) used the SRS successfully to predict competitive sorption of Cd, Ni, and Zn in a Greek vertic xerochrept soil. They found Zn was strongly retained and competition suppressed the sorption of the three metals.

In the example we present here, three surface soils with contrasting properties were chosen for study: a Webster Loam, Windsor sand, and Olivier loam. Webster soil has a pH of 6.92 and cation exchange capacity (CEC) of 27.0 cmol kg^{-1}. For Windsor sand, the pH is 6.11 and the CEC is 2.0 cmol kg^{-1}, and for Olivier loam they are pH of 5.8 and CEC of 8.6 cmol kg^{-1}. Batch adsorption of Ni and Cd in single and binary Ni-Cd systems was carried out in the traditional methods (Selim and Amacher, 1997). Different molar ratios of Ni/Cd for a wide concentration range were applied to investigate competitive Cd and Ni in all soils. Sorption isotherms for single ions as well as binary systems were modeled using the Freundlich and competitive approaches. A wide range of concentrations of the competing ions is necessary to delineate the adsorption characteristics for different heavy metals as well for modeling multicomponent competitive systems.

Sorption isotherms Ni and Cd are shown in Figure 7.13 for Olivier, Windsor, and Webster soils. These isotherms exhibit highly nonlinear behavior indicative of strong affinities at low heavy metal concentrations. For all three soils, the overall shape of the isotherms suggests some similarities in sorption mechanisms of the two cations. The Freundlich approach was used to describe both Ni and Cd isotherms. For both Ni and Cd the Freundlich well described the isotherms with coefficients of correlation (R^2) ranging from

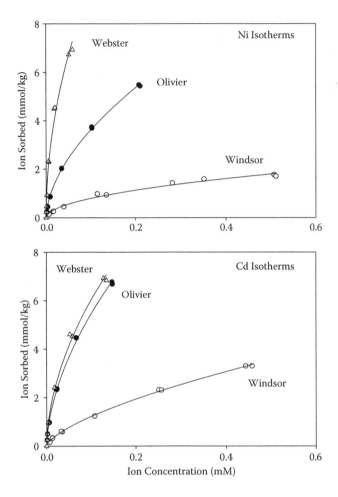

FIGURE 7.13
Adsorption isotherms for Ni (top) and Cd (bottom) for Windsor, Olivier, and Webster soils. Solid curves are Freundlich model calculations.

0.982 to 0.999 (Liao and Selim, 2010). The dimensionless parameter n may be regarded as a representation of energy distribution of heterogeneous adsorption sites for solute retention by matrix surfaces (Sheindorf, Rebhun, and Sheintuch, 1981). Nonlinearity and competition are often regarded as characteristics of site-specific adsorption processes. Here adsorption occurs preferentially at the sites with highest adsorption affinities and sites with lower adsorption potential are occupied with increasing concentration. The n values for Windsor, Olivier, and Webster soils are 0.64, 0.57, and 0.55 for Cd and 0.50, 0.56, and 0.52 for Ni, respectively. These n values were within a narrow range (0.50–0.64) for all three soils and reflect the observed similarities of the overall shape of both Ni and Cd sorption isotherms as shown in

Figure 7.13. Moreover, the shape of these isotherms depicts an L-type curve as described by Sposito (1984). These *n* values are within the range of values of those reported earlier by Buchter et al. (1989); 0.57–0.78 for Cd and 0.65–0.74 for Ni. A comparison of the adsorption isotherms indicates that for both Ni and Cd sorption affinities follows the sequence; Windsor < Olivier < Webster soil (Figure 7.11). This is also illustrated by the respective *K* values for Cd; 5.62, 24.59, and 26.78 L kg^{-1} and for Ni; 2.55, 13.30, and 37.57 L kg^{-1}. The work of Gomes, Fontes, and da Silva (2001), among others, indicates that Cd and Ni adsorption by a number of soils are correlated with CEC. Papini, Saurini, and Bianchi (2004) reported that Cd and Ni adsorption was largely due to cation-exchange reaction on an Italian red soil.

Results of competitive Ni sorption in the presence of a range of Cd concentrations are given in Figure 7.14 for all three soils. Here the amount of Ni sorbed (mmol per kg soil) is presented versus input concentration of the competing Cd ion for two initial Ni concentration, 0.025 mM (Figure 7.14,

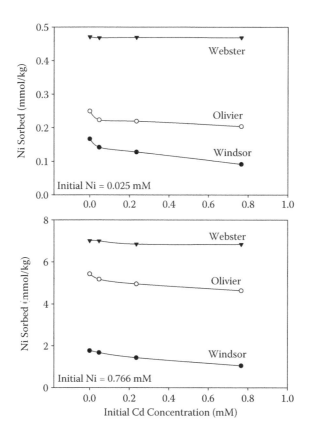

FIGURE 7.14
Competitive sorption of Ni in the presence of Cd for Windsor, Olivier, and Webster soils. Initial Ni concentrations were 0.025 mM (top) and 0.766 mM (bottom).

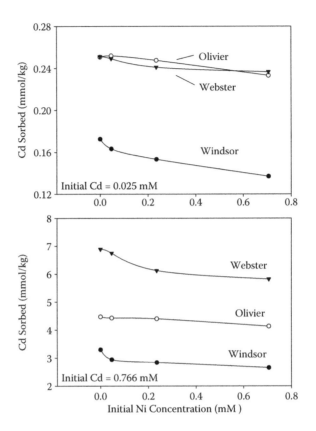

FIGURE 7.15

Competitive sorption of Cd in the presence of Ni for Windsor, Olivier, and Webster soils. Initial Cd concentrations were 0.025 mM (top) and 0.766 mM (bottom).

top) and 0.766 mM (Figure 7.14, bottom). These results indicate that Ni sorption decreased as the competing Cd concentration increased. In Figure 7.15, results are shown for Cd sorption in the presence of a range of Ni concentrations for all three soils. Here Cd adsorption decreased with increasing Ni concentrations. Moreover, the extent of the decrease in Ni or Cd sorption in this competitive systems was dissimilar among the three soils. For the two acidic soils (Windsor and Olivier), Ni adsorption decreased substantially with increasing Cd concentration in comparison to Webster, the neutral soil. This finding was consistent for both initial Ni concentrations (0.025 and 0.766 mM—see Figure 7.14). The amount of Ni sorbed in the presence of 0.766 mM Cd was reduced by 0.5%, 18%, and 45% for Webster, Olivier, and Windsor soil, respectively. When 0.766 mM Ni was present, sorbed Cd was reduced by 20%, 7.6%, and 15% for Windsor, Olivier, and Webster soil, respectively.

Metal ion competition is presented in the traditional manner as isotherms and is given in Figures 7.16 and 7.17. These isotherms were described using the Freundlich model in a similar manner to those for a single ion. The extent of nonlinearity of Ni and Cd isotherms is depicted by the dimensionless parameter n and was not influenced by input concentration of the competing ion. Thus in a competitive system, the parameter n did not exhibit appreciable changes for both metal ions investigated. In contrast, K values exhibited a decrease of sorption as the concentration of the competing ion increased, and the extent of such a decrease was dissimilar for the three soils. For Windsor and Olivier, Ni adsorption decreased significantly over the entire range of concentrations of the competing ion (Cd). However, Cd adsorption was less affected by the competing Ni ions for both soils. For the neutral Webster soil, Ni was not appreciably affected by the presence of Cd, especially at low Ni concentrations. This may be because, for a single-component system, Ni adsorption was much stronger than Cd for Webster soil as discussed above. Another explanation of the competitive Ni sorption behavior is perhaps due to Ni-LDH precipitates, which may be considered irreversible on soils and minerals (Voegelin and Kretzschmar, 2005). This process may lead to significant long-term stabilization of the metal within the soil profile (Ford et al., 1999). In acidic soils, Ni and Cd are both weakly bonded to soil particle surfaces and mainly forms outer spheres, which are available for cation exchange. However, for the neutral Webster soil, Ni sorption may include a fraction of inner-sphere complexation or Ni-LDH precipitates, both of which are perhaps not available for competition via cation exchange.

Based on Sheindorf-Rebhun-Sheintuch (SRS) model predictions, the estimated $\alpha_{Ni\text{-}Cd}$ for Ni adsorption, in the presence of Cd, was larger than 1 for Windsor and Olivier soils, indicating noticeable decrease of Ni in the presence of Cd. In contrast, $\alpha_{Ni\text{-}Cd}$ for Ni adsorption on Webster soil was less than 1, which is indicative of small influence of competing Cd ions. These results are in agreement with the competitive sorption reported by Antoniadis and Tsadilas (2007). Such small $\alpha_{Ni\text{-}Cd}$ implies that Ni adsorption in Webster soil was least affected in a competitive Ni-Cd system in comparison to the other two soils. Moreover, the estimated $\alpha_{Cd\text{-}Ni}$ for Cd adsorption was 0.61 for Windsor and 0.82 for Olivier, whereas the competitive coefficient of Cd/Ni was 4.00 for Webster soil. Although the SRS equation may be regarded as a multicomponent model and does not imply certain reaction mechanisms, differences of competitive sorption between the neutral and the two acidic soils were illustrated based on the SRS models' competitive selectivity parameters. In fact, Roy, Hassett, and Griffin (1986a) suggested that the SRS parameters could be used to describe the degree of the competition under specific experimental conditions. Calculated results using the estimated $\alpha_{Ni\text{-}Cd}$ are given in Figures 7.18 and 7.19 and illustrate the capability of the SRS model to describe experimental data for competitive adsorption of Ni and Cd. An *F*-test indicated that there was no statistical difference between our experimental results and SRS model calculations (at the 95%

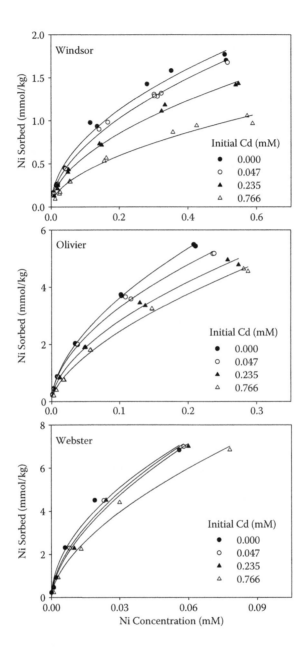

FIGURE 7.16

Competitive adsorption isotherms for Ni in the presence of different concentrations of Cd. Solid curves are Freundlich model calculations.

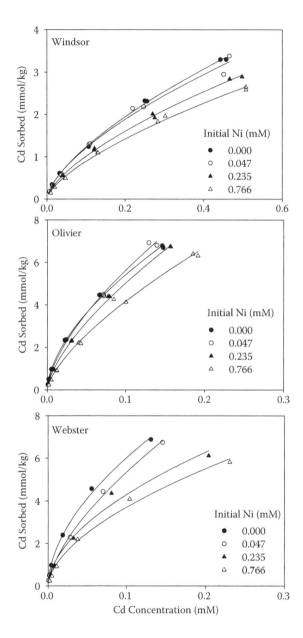

FIGURE 7.17
Competitive adsorption isotherms for Cd in the presence of different concentrations of Ni.
Solid curves are Freundlich model calculations.

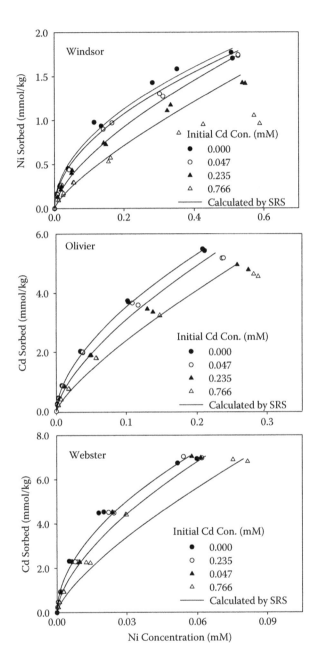

FIGURE 7.18
Competitive adsorption isotherms for Ni in the presence of different concentrations of Cd. Solid curves are SRS model calculations.

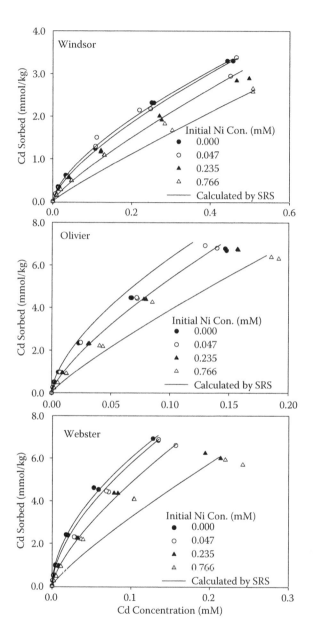

FIGURE 7.19

Competitive adsorption isotherms for Cd in the presence of different concentrations of Ni. Solid curves are SRS model calculations.

confidence level). Based on these calculations, the SRS model was capable of quantifying competitive adsorption for Ni and Cd. However, for both Ni and Cd, the SRS model deviated considerably from experimental data for high concentrations of the competing ions. This finding is consistent with the application of SRS made earlier by Gutierrez and Fuentes (1993) and illustrates the need for model improvement to better describe competitive adsorption of heavy metals over the entire range of concentrations.

7.8.2 Zinc–Phosphate

Zinc availability and mobility in soils may be controlled by several interactions with the soil-water environment. Primary sources of Zn contamination include mining, smelting, and other industrial as well as anthropogenic factors (Adriano, 2001). A secondary source of Zn is through phosphate fertilizers, which often contain traces of heavy metals such as Cd, Cu, Mn, Ni, Pb, and Zn. Zinc is also an essential micronutrient for plants and animals (Adriano, 2001). The understanding of the complex interactions of Zn in the environment is a prerequisite in the effort to predict their behavior in the vadose zone. It is well accepted that several factors influence Zn adsorption, desorption, and equilibrium between the solid and solution phases. These factors include soil pH, clay content, organic matter (OM), cation exchange capacity (CEC), and Fe/Al oxides (Gaudalix and Pardo, 1995), among which, soil pH is one of the most important factors (Barrow, 1987). Zn sorption increases and Zn desorption reduces with increased pH (Rupa and Tomar, 1999; Tagwira, Piha, and Mugwira, 1993). This may be because increasing pH increases the negative charge of variable-charge soil for Zn adsorption (Saeed and Fox, 1979). On the other hand, pH affects Zn hydrolysis which would be preferentially sorbed on soil surface (Bolland, Posner and Quirk, 1977).

Over the past two decades, phosphate has been observed to increase Zn adsorption and decrease Zn desorption in soils (Agbenin, 1998; Rupa and Tomar, 1999). Xie and Mackenzie (1989) reported that P sorption increased soil CEC, resulting in increased Zn adsorption on three different soils. Saeed and Fox (1979) reported that increased negative charge due to P sorption was responsible for the observed increase in Zn sorption. Xie and Mackenzie (1989) found phosphate sorption enhanced the correlation between Zn sorption and soil OM and Fe content and postulated that enhanced Zn sorption may be a consequence of either an increase in negative surface charge of soil particles, creation of specific sites, or precipitation of hoepite. Ahumada, Bustamnte, and Schalscha (1997), Pardo (1999), and Sarret et al. (2002) observed significant increases in Zn fractions associated with organic matter and Fe/Al oxides in the presence of P. Rupa and Tomar (1999) investigated Zn desorption kinetics as influenced by phosphate on alfisol, oxisol, and vertisol. They tested several kinetic models and found that Zn release was best described by the Elovich equation. A primary limitation of simple models such as

Elovich is that different rates of reaction are needed to describe Zn adsorption from Zn desorption or release. Multireaction kinetic models are capable, given only one set of reactions rate parameters, of describing adsorption and desorption or release and have been successfully used to describe kinetic behavior for several heavy metals and contaminants (Amacher, Selim, and Iskandar, 1988; Darland and Inskeep, 1997; Barnett et al., 2000). None of the multireaction-type models have been tested for their capability to describe Zn reactivity in the complex soil environment in the presence of P.

Similar to the previous case study for Ni-Cd, three surface soils with contrasting properties were chosen for this study: a Webster Loam, Windsor sand, and Olivier loam. Adsorption of Zn was studied using the batch method as described by Selim and Amacher (1997). The duration of adsorption was 1 day. Release or desorption commenced following the 1 day adsorption step using sequential or successive dilutions. To study the influence of P on Zn adsorption as well as release, the Zn batch experiments were also carried out where different levels of P concentrations were added in the solutions.

A family of Zn isotherms is represented in Figure 7.20 for each soil, where different initial P concentrations were added (from 0 to 100 mg L^{-1}) (see Zhao and Selim, 2010). Distinct differences in the amount of Zn sorbed among the different soils were observed. Highest sorption was observed for the neutral Webster soil. Strong retention of Zn is observed as the soil pH increased (Harter and Naidu, 2001: Barrow, 1987). Moreover, Webster soil is a fine loamy Haplaquoll with 3.7% calcium carbonates and the presence of carbonates enhances Zn sorption in soils (see Mesquita and Vieira e Silva, 1996). The Olivier and Windsor soils exhibited lower sorption capacities where Windsor soil showed least sorption. Hinz and Selim (1994) reported stronger retention for Zn by Olivier compared to Windsor which may be due to higher CEC caused predominately by smectitic clays. In contrast, Windsor is an Entisol and contains parent material that has not been completely weathered to secondary minerals and hence lower sorption capacity for Zn. Increased CEC results in more negatively charged sites to adsorb Zn (Kurdi and Doner, 1983; Cavallaro and McBride, 1984).

The influence of P on increased Zn sorption was clearly manifested in the isotherms shown in Figure 7.20 where similar trends were observed for all soils. Saeed and Fox (1979) showed that P fertilization increased Zn adsorption by soils from Hawaii that contained colloids predominantly of the variable-charge type. Their results support the hypothesis that phosphate additions to soils increased Zinc adsorption by increasing the negative charges on iron and aluminum oxides. Wang and Harrell (2005) reported that zinc sorption was enhanced by $H_2PO_4^-$ as opposed to Cl^- or NO_3^- in acid soils. The overwhelming increase of total Zn sorption with P addition was consistent with other studies (Agbenin 1998; Pardo, 1999; Rupa and Tomar, 1999; Xie and Mackenzie, 1989). Moreover, the influence of P on increased Zn sorption for Windsor and Webster soils was much greater than for Olivier soil.

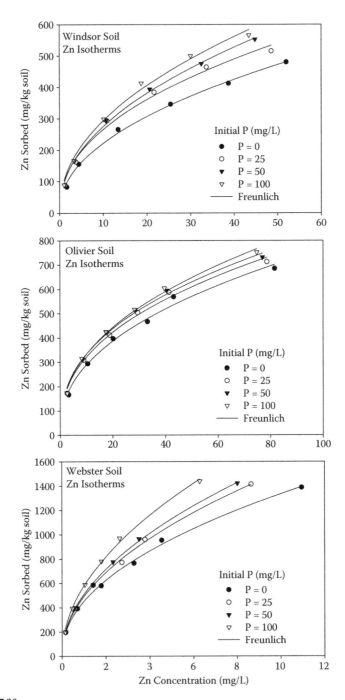

FIGURE 7.20
Zinc adsorption isotherms on different soils after 24 h sorption in the presence of different P concentrations. Solid curves are simulations using the Freundlich equilibrium equation.

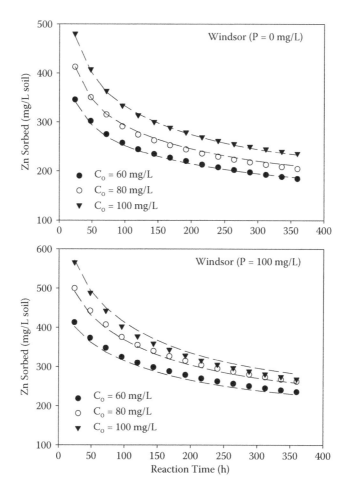

FIGURE 7.21
Zn concentration in Windsor soil versus time during adsorption and desorption for various initial Zn concentrations. Dashed curves are multireaction model (MRM) simulations.

Figures 7.21 and 7.22 present the amount of Zn sorbed versus time to illustrate the kinetics of Zn desorption for the various initial concentrations (C_o) used. As illustrated in the figures, Zn desorption exhibited strong time-dependent behavior as depicted by the continued decrease of the amount sorbed with time. For Windsor and Olivier soils, the rate of Zn desorption was initially rapid and followed by gradual or slow reactions. In contrast, for Webster slow release reactions for Zn were dominant. The results indicate that a fraction of Zn was weakly sorbed by Windsor and Olivier soils via ion exchange or outer-sphere surface complexation. In contrast, Zn was strongly sorbed in Webster soil and bound via inner-sphere surface complexation. Consistent with these

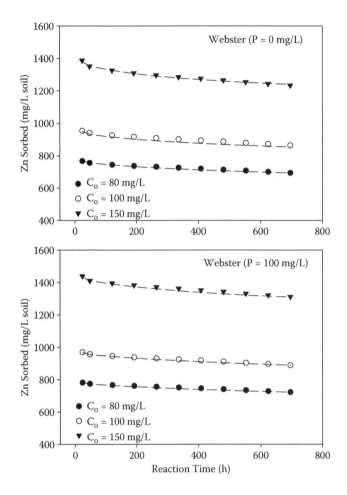

FIGURE 7.22
Zn concentration in Webster soil versus time during adsorption and desorption for various initial Zn concentrations. Dashed curves are multireaction model (MRM) simulations.

results, Zn desorption kinetic results reported by Rupa and Tomar (1999) showed that for increasing level of applied P to 20 and 40 mg/kg subsequent decrease in Zn desorption by 20% to 31% and 39% to 53% were observed.

As previously reported the P-induced Zn sorption may be related to increased surface negative charges and/or creation of specific sorption sites as a result of P sorption on OM, Fe/Al oxides and clays (Barrow, 1987; Bolland, Posner, and Quirk, 1977; Xie and Mackenzie, 1989). Xie and Mackenzie (1989) postulated that sorption of P on OM and oxides can form inner-sphere surface complexes that increase negative charges, resulting in increased surface Zn retention. Kuo and McNeal (1984) proposed that the added P act as a "bridge" between soil surfaces and sorbed Zn, or sorbed P alters the soil properties sufficiently to affect Zn sorption. The latter was also proposed by Tagwira, Piha, and Mugwira (1993). Thus, P application can affect Zn sorption resulting in increased specific sorption sites for Zn and subsequently reducing Zn release or desorption. Furthermore, P-induced Zn retention may be due in part to precipitation or co-precipitation involving the formation of solid-solution of $ZnHPO_4$ as an intermediate product (Agbenin, 1998). Al-P precipitation products from reactions of P with Al-OM complexes may form new sites to retain Zn on soil organic matter. Thus, P sorption on the surface increased Zn sorption and restricted Zn desorption, depending on soil pH, surface complexation, and soil precipitation.

Desorption or release following adsorption is presented as isotherms in the traditional manner in Figures 7.23 and 7.24 for both soils. Distinct discrepancies between adsorption and successive desorption isotherms are clearly observed in the figures and indicate considerable hysteresis for Zn release, the extent of which varied among the three soils. This observed hysteresis was not surprising in view of the kinetic behavior of Zn and is indicative of the nonequilibrium behavior of Zn retention mechanisms. Significant irreversibility of Zn sorbed on mineral surfaces and soils has been reported extensively in the literature, and this observed hysteresis may be due in part to slow diffusion and kinetic ion exchange, as well as irreversible mechanisms.

Hysteresis results shown in Figures 7.23 and 7.24 illustrate the extent of kinetics during release of Zn in the three soils. Webster exhibited limited kinetics or very slow release where desorption isotherms exhibited little release over time for all initial Zn concentrations considered. In contrast, for the two acidic soils Zn release as a percentage of the total sorbed was 47% to 51% and 42% to 49% for Windsor and Olivier soil, respectively. For the neutral Webster, only 9% to 11% of the sorbed Zn was released into the soil solution after 28 d of successive desorptions. This is due to its much higher content of clay with smectite as the predominating mineral, organic matter, Fe/Al oxides, and CEC. Soils with high clay, OM, Fe/Al content, and CEC have more available reactive sites and increased specific sorption and reduced Zn desorption (Xie and MacKenzie, 1989). Furthermore, at low initial Zn concentrations, lower proportions of Zn were desorbed, indicating high affinity for Zn in soils, whereas at high concentration, the percentage of Zn desorption increased, indicating lower Zn affinity. Although higher Zn amounts were sorbed as a result of the presence of P, similar release curves were obtained for all P concentrations (Figures 7.23 and 7.24). Thus, the kinetics of Zn retention was not altered by the presence of increased levels of P. However, the application of P resulted in reduced desorption of Zn for all three soils. This is perhaps because the added P increased specific sorption sites on OM and Fe/Al oxides surfaces where Zn was tightly sorbed. Moreover, the effect of P on Zn sorption by Webster soil was more pronounced compared to the acidic soils. This is likely due to the high clay, OM, Fe/Al content, and CEC for Webster soil as discussed above.

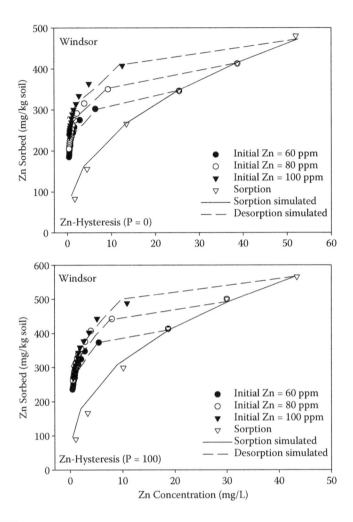

FIGURE 7.23
Adsorption and desorption isotherms for Windsor soil with different P concentration. The solid and dashed curves depict results of multireaction (MRM) model simulations for sorption and desorption.

In order to assess the mobility of Zn in soils, column transport experiments were carried out using the miscible displacement technique. Transport of Zn in soils was quantified using miscible displacement techniques in water-saturated soil columns. A Zn pulse solution with or without P was applied to each column. In Figure 7.25, Zn BTCs are presented for a Windsor soil column that received a pulse of Zn alone (column 1). In contrast, in Figure 7.26, results show BTCs where a mixed pulse of P and Zn was introduced (column 2). A comparison of the two Zn BTCs clearly shows increased retention of Zn sorption in the presence of P (see Figure 7.27). In fact after some 120 pore

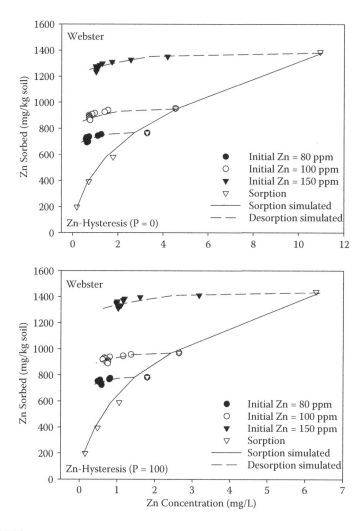

FIGURE 7.24
Adsorption and desorption isotherms for Webster soil with different P concentration. The solid and dashed curves depict results of multireaction (MRM) model simulations for sorption and desorption.

volumes, 55% of applied Zn was retained in the presence of P compared to 31% retention without P. Overall Zn concentration in the effluent decreased in the presence of P without a change in the relative position of the BTC; for example, arrival of the Zn front in the effluent solution. Such behavior is indicative of increased irreversible Zn retention in the presence of P. It is clear that the presence of P resulted in a lower peak concentration and less overall Zn recovery in the effluent when compared to column 1 BTCs where P was absent. The transport data are consistent with retention and clearly indicate that the presence of P increases the amount of Zn sorbed. Moreover,

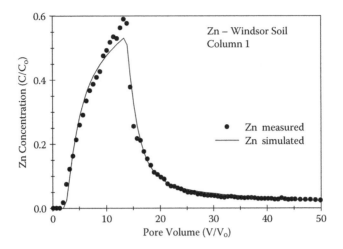

FIGURE 7.25
Breakthrough curve for Zn from Windsor soil column 1. The solid curve is based on multireaction model simulation.

all BTCs were successfully described by the multireaction model presented above.

7.8.3 Arsenate–Phosphate

Arsenic (As) is a highly toxic element widely present in trace amounts in soil, water, and plants. Adsorption to soils and sediments is the major pathway

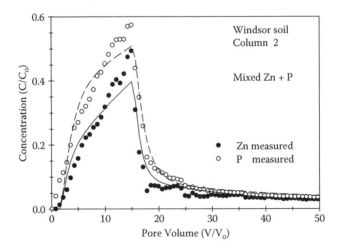

FIGURE 7.26
Breakthrough curves for Zn and P from Windsor soil column 2. Solid and dashed curves are based on multireaction model simulations.

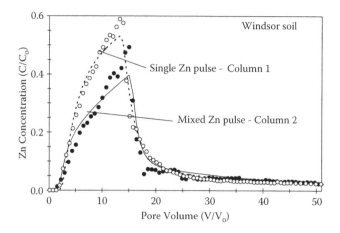

FIGURE 7.27
Breakthrough curves for Zn from a single pulse (column 1) and a mixed (P + Zn) pulse (column 2) in Windsor soil.

of attenuating arsenic bioavailability and toxicity in natural environments (Smedley and Kinniburgh, 2002). Phosphate (P) anion has similar chemical properties to arsenate [As(V)] and forms similar types of inner-sphere surface complex with metal oxides. The competition between As(V) and P for adsorption sites has the potential of increasing arsenic mobility and bioavailability in soil environments (Woolson, Axley, and Kearney, 1973; Peryea, 1991).

The competitive adsorption of As(V) and P can be affected by a wide range of factors, such as surface properties of the adsorbent, concentration and molar ratio of As(V) to P, solution pH, and residence time (Hingston, Posner, and Quirk, 1971; Violante and Pigna, 2002; Liu, De Cristofaro, and Violante, 2001; Jain and Loeppert, 2000; Manning and Goldberg, 1996a, 1996b; Zhao and Stanforth, 2001). Hingston, Posner, and Quirk (1971) proposed two types of adsorption sites on mineral surfaces. The first type is available for both As(V) and P where competition takes place while the second type is specifically available for either As(V) or P. Results from studies on single ion sorption showed that As(V) and P sorption on Fe and Al oxides were somewhat similar (Manning and Goldberg, 1996a, 1996b). However, when added simultaneously in equal molarities, Violante and Pigna (2002) reported that metal oxides and phyllosilicates rich in Fe were more effective in adsorbing As(V) than P, while more P was adsorbed than As(V) on minerals rich in Al. Adsorption studies on soils often reveal that P is preferentially adsorbed than As(V) whether added separately or added simultaneously in equal molar ratios (Roy, Hassett, and Griffin, 1986a, 1986b; Smith, Naidu, and Alston, 2002; De Brouwere, Smolders, and Merckx, 2004). These studies also demonstrated that amounts of As(V) sorbed on minerals and soils exhibited a decrease with increasing additions of P in solution. Moreover, the sequence of addition might significantly affect the competition between As(V) and P (Liu, De Cristofaro, and Violante, 2001; Zhao and Stanforth, 2001). Liu, De Cristofaro, and Violante (2001) found that when added sequentially [As(V) before P versus P before As(V)], more P was replaced by As(V) on goethite than vice versa. Due to the similar dissociation constants of phosphate ($pK_a^1 = 2.23$, $pK_a^2 = 7.2$, $pK_a^3 = 12.3$) and arsenate ($pK_a^1 = 2.20$, $pK_a^2 = 6.97$, $pK_a^3 = 11.53$), adsorption of both anions decreases with increasing pH (Jain and Loeppert, 2000). When added simultaneously, Jain and Loeppert (2000) reported that the effect of P on As(V) adsorption on ferrihydrite was greater at high pH than at low pH.

In most studies dealing with competitive adsorption, equilibrium conditions are often assumed. However, several experiments demonstrated adsorption of As(V) (Fuller, Davis, and Waychunas, 1993; Grossl et al., 1997; Raven, Jain, and Loeppert, 1998; Waltham and Eick, 2002; Zhang and Selim, 2008) and P (Barrow, 1992; Wilson, Rhoton, and Selim, 2004) on minerals and soils are both time dependent.

In the case study presented here, As adsorption as well as desorption results are presented in the presence of different P concentration (see Zhang and Selim, 2008). Furthermore, transport results from soil columns that received As alone and columns that received a pulse having As and P (see Zhang and Selim, 2007). In the subsequent sections, these two treatments are referred to as single pulse and mixed pulse, respectively.

Amount of As(V) sorption in the presence of P is depicted in Figure 7.28, which indicates that As(V) adsorption decreased substantially with increasing P concentrations. Competition for specific adsorption sites is likely the major cause for the observed competitive effect between As(V) and P shown

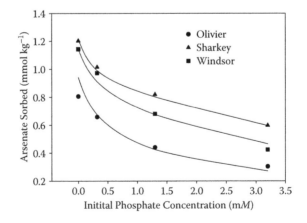

FIGURE 7.28
Competitive sorption between arsenate and phosphate at 24 h of reactions for Olivier, Sharkey, and Windsor soils. The initial concentration of arsenate was 0.13 mM.

in Figure 7.28. Formation of surface complexation between Fe/Al (hydro) oxides and As(V)/P restricted the accessibility of those surface sites for further adsorption (Hingston, Posner, and Quirk, 1971; Smith, Naidu, and Alston, 2002). Because adsorption of both anions takes place preferentially on high affinity sites at low surface coverage, the competition is expected to be greatest at low As(V) and P concentrations, which is consistent with observations.

The SRS equation (7.15) was employed to quantify competitive adsorption of As(V) and P. Specifically, the Freundlich K_F and N were taken from the single-anion adsorption data and utilized to obtain the competitive coefficients α_{ij} by fitting the competitive adsorption data to Equation 7.15 using nonlinear least square optimization. We should emphasize that the SRS equation should only be regarded as an empirical model and the conformity of this equation does not imply certain reaction mechanisms. In their original paper, Sheindorf, Rebhun, and Sheintuch (1981) defined α_{ij} as symmetrical values, that is, $a_{ij} = 1/a_{ji}$. However, Roy, Hassett, and Griffin (1986a, 1986b) suggested that the coefficients should be regarded as empirical values describing the degree of competition under specific experimental conditions. Furthermore, Barrow, Cartes, and Mora (2005) used nonlinear curve fitting to determine the competitive coefficients between As(V) and P and they found that the coefficients were not symmetrical.

Results from the kinetic batch experiments are presented in Figures 7.29 and 7.30 in order to illustrate the competitive sorption kinetics between As(V) and P by the soils. The extent of As(V) sorbed by all soils were significantly reduced as concentrations of P in the applied solution increased. Moreover, both As(V) and P exhibited strongly time-dependent adsorption behavior, which is depicted by the continued decrease of concentration with reaction time. Observed retention kinetics of As(V) and P in Figures 7.29 and 7.30 is likely due to the heterogeneity of the soil surface where multiple chemical and physical processes take place. Chemical reaction rates of surface complexation between anions and metal oxides are considered rapid. Using a pressure jump relaxation technique, Grossl et al. (1997) calculated a kinetic rate constant of $10^{6.3}$ s^{-1} for the formation of a monodentate inner-sphere surface complex on goethite surface. In addition, a forward rate constant of 15 s^{-1} was associated with succeeding reaction for the formation of a bidentate mononuclear surface complex. Because of their rapid reaction rates, surface complexation is not a rate-limiting step of As(V) and P adsorption in soils. However, different types of surface complexes (e.g., monodentate, bidentate, mononuclear, binuclear) can be formed on oxide surfaces at high or low surface coverage. This heterogeneity of sorption sites may contribute to adsorption kinetics observed in most experiments, that is, where sorption takes place preferentially on high affinity sites and followed by slow sorption to sites of low sorption affinity.

Recent adsorption studies suggested that surface precipitation, that is, three-dimensional growth of a particular surface phase, may occur for both

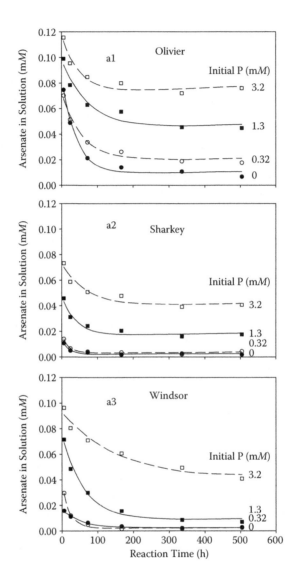

FIGURE 7.29
Arsenate concentrations in solution versus time in the presence of various concentrations of phosphate. The initial concentration of As(V) was 0.13 mM.

As(V) and P (Zhao and Stanforth, 2001; Pigna, Krishnamurti, and Violante, 2006; Jia et al., 2006). The development of surface precipitate is a slow process involving multiple reaction steps and explains in part the slow As(V) and P retention kinetics. The theory of surface precipitation suggests that anions absorbed on mineral surfaces attract dissolved Fe or Al. Adsorbed

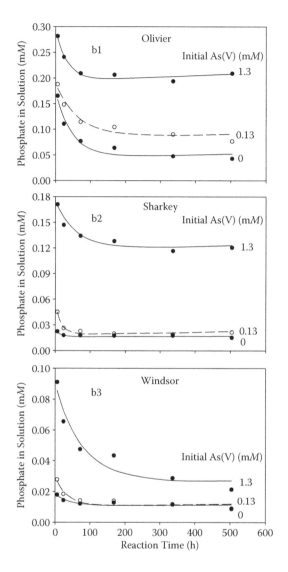

FIGURE 7.30
Phosphate concentrations in solution with time in the presence of various concentrations of arsenate. The initial concentration of P was 0.32 mM.

Fe or Al in turn adsorbs more anions and results in a multilayer adsorption. Zhao and Stanforth (2001) suggested the slow buildup of surface precipitate as the mechanisms of irreversible As(V) and P retention on goethite. More recently, the x-ray diffraction and Raman spectroscopy results of Jia et al. (2006) confirmed the formation of poorly crystalline ferric As(V) surface precipitate on ferrihydrite under high As/Fe molar ratio, low pH, and long reaction time.

Diffusion of As(V) and P to reaction sites within the soil matrix was also proposed as the explanation of the time-dependent adsorption (Fuller, Davis, and Waychunas, 1993; Raven, Jain, and Loeppert, 1998). A two-phase process was generally assumed for diffusion-controlled adsorption, with the reaction occurring instantly on liquid-mineral interfaces during the first phase, whereas slow penetration or intraparticle diffusion is responsible for the second phase. The pore space diffusion model has been employed by Fuller, Davis, and Waychunas (1993) and Raven, Jain, and Loeppert (1998) to describe the slow sorption of As(V) on ferrihydrite. For heterogeneous soil systems, the complex network of macro- and micro-pores may further limit the access of solute to the adsorption sites and cause the time-dependent adsorption.

Transport results of As(V) are presented by the BTCs in Figures 7.31 and 7.32. Each soil column received two consecutive As pulses. The BTCs indicate extensive As(V) retention during transport in both soils. After two As(V) pulse applications and subsequent leaching by arsenic-free solution for more than 20 pore volumes, As(V) mass recoveries in the effluent were 82.1% and 72.5% of that applied for Olivier and Windsor soil, respectively. The BTCs were asymmetric, showing excessive tailing during leaching.

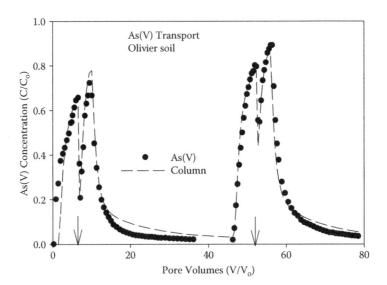

FIGURE 7.31

Experimental As(V) breakthrough curves in Olivier soil without addition of P. Solid curves are single-component multireaction model (MRM) predictions using batch kinetic parameters. The dashed curves depict MRM results based on nonlinear optimization. Arrows indicate pore volumes when flow interruptions occurred.

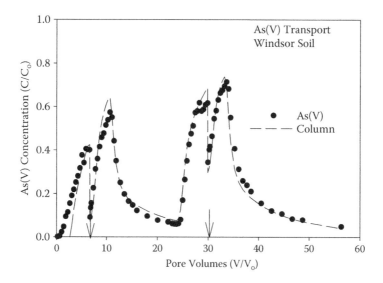

FIGURE 7.32

Experimental As(V) breakthrough curves in Windsor soil without addition of P. Solid curves are single-component multireaction model (MRM) predictions using batch kinetic parameters. The dashed curves depict MRM results based on nonlinear optimization. Arrows indicate pore volumes when flow interruptions occurred.

The transport of As(V) in the presence of P is illustrated by the BTCs shown in Figures 7.33 and 7.34. Similar to As(V), BTCs for P exhibited extensive asymmetry. Nonequilibrium conditions appear dominant as indicated by the sharp drop in P concentration as a result of flow interruption (or stop flow). In order to describe the competitive transport of As(V) and P, a model was developed that accounts for equilibrium and kinetics mechanisms. Specifically, the equilibrium and kinetic adsorption equations were modified in a way similar to the competitive SRS model. The single-species simulation results were further tested with the analytical solution for the two-site nonequilibrium transport model provided by CXTFIT (Toride, Leij, and van Genuchten, 1995).

7.8.4 Vanidate–Phosphate

Vanadium is a ubiquitous trace element in the environment and is an essential trace element for living organisms, but in excessive amounts is harmful to humans, animals, and plants (Crans et al. 2004). Vanadium acts as a growth-promoting factor and participates in fixation and accumulation of nitrogen in plants, whereas high concentrations of vanadium may reduce their productivity (Underwood, 1977). In soils, vanadium is derived from parental rocks and deposits.

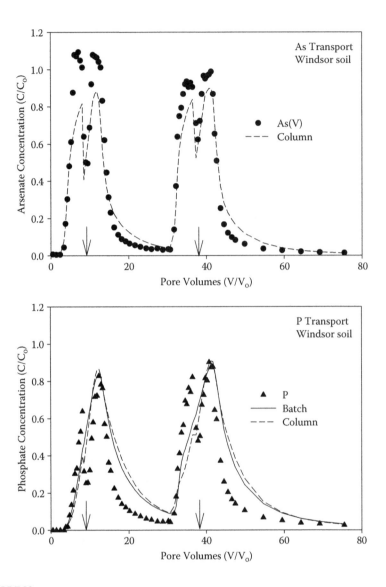

FIGURE 7.33
Phosphate concentrations versus reaction time with the presence of various concentrations of arsenate for Olivier and Windsor soils. The initial P concentration was 0.32 mmol L[1]. Symbols are for different initial As(V) concentrations of 0, 0.13, and 1.29 mmol L[1]. Solid and dashed curves are multicomponent, multireaction model simulations.

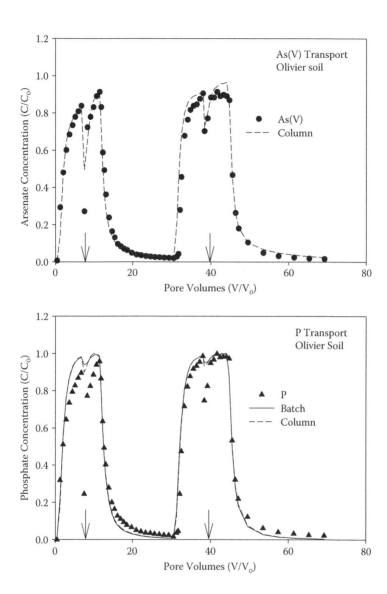

FIGURE 7.34

Experimental As(V) and P breakthrough curves in Windsor soil (column 6). Solid curves are multicomponent, multireaction model (MCMRM) predictions using batch kinetic parameters. The dashed curves are MCMRM simulations using kinetic parameters obtained from single-component As(V) transport experiments. Arrows indicate pore volumes when flow interruptions occurred.

Environmentally, the occurrence of vanadium in petroleum and coal is of high significance because fuels constitute major sources of vanadium emissions to the atmosphere (Dechaine and Gray, 2010). A large fraction of the vanadium-rich atmospheric particles may enter the soil environment as particulate fallout or dissolved in rain. Vanadate and vanadyl ions are versatile at forming complexes that inhibit or stimulate activity of many enzymes by specific mechanisms.

Here we present vanadium transport and adsorption results for two soils with contrasting properties: a Sharkey clay (very fine, montmorillonitic, nonacid, Vertic Haplaquept) and Cecil soil ((clayey, Kaolinitic, thermic Typic Kanhapludult). The Sharkey soil has an organic matter content of 1.41%o and a pH of 5.9, whereas Cecil soil has an organic matter content of 0.74%o and a pH of 5.6. Miscible displacement experiments were carried out using soil columns under saturated condition and state-flux was slowly saturated upward with 0.01 N NH_4Cl background solution for 4 d prior to V pulse application. A vanadium pulse of 100 mg L^{-1} of NH_4VO_3 in 0.01 N NH_4Cl was applied to each column and was followed by 0.01 N NH_4Cl solution. The effluent solution from each column was collected using a fraction collector. To assess the impact of the presence of P on vanadium mobility in the soil columns, 100 mg L^{-1} of P as -$NH_4H_2PO_4$ was mixed with the vanadium pulse solution. Moreover, periods of stop-flow (or flow interruption) of 2 d duration were implemented to test whether equilibrium or kinetic processes are the dominant transport mechanism.

Vanadium adsorption by the two soils represents contrasting behavior with regard to intensity as well as kinetics. Sharkey soil exhibited stronger affinity for vanadium than Cecil soil. In addition, Cecil soil, which is predominantly kaolinitic, exhibited extremely limited kinetic reaction (see Figures 7.35 and 7.36). When compared to 1 d retention, after 4 d of reaction an increase of only 30% was observed. In contrast for Sharkey soil, which is predominantly montmorillonitic, there was a threefold increase in vanadium adsorption when compared to 1 d sorption. The K_d values for Cecil soil were 1.70 and 2.31 L/kg, after 1 and 4 d sorption, respectively. The corresponding K_d values for Sharkey soil were 73.03 and 360.91 L/kg, after 1 and 21 d sorption, respectively. The presence of P resulting in decreasing the amount of V sorbed varied among the two soils. In fact, the impact of P on reduced vanadium sorption was greatest for Sharkey soil, where a onefold decrease was observed. In contrast, only 50% decrease in vanadium sorption was observed for Cecil soil (Figures 7.35 and 7.36).

The mobility of vanadium (V) as well as phosphorus (P) is presented by the BTCs presented in Figures 7.37 and 7.38. These BTC's represent different column scenarios. In column 1, a pulse of vanadium was followed by a pulse of phosphorus, which was subsequently followed by a pulse of a mixed solution of P plus V. In Figure 7.37, the sequence of pulse applications was P followed by a pulse of V and subsequently by a mixed pulse. For column 2 shown in Figure 7.38, the soil column received three consecutive pulses of

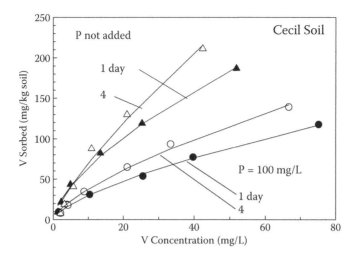

FIGURE 7.35
Vanadium isotherms after 1 and 4 d retention for Cecil soil. Isotherms are also presented in the presence of 100 mg of P in the soil solution.

mixed solution (P plus V). The BTCs clearly indicate higher affinity for V over P for all cases presented here. This was evident when three consecutive pulses were introduced in the Cecil soil column, as illustrated in Figure 7.38. For the case shown in Figure 7.37, both P and V peaks represent the input pulse solution during all three pulses. Based on peak concentration and

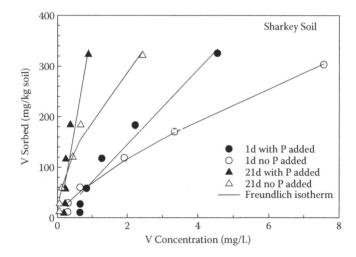

FIGURE 7.36
Vanadium isotherms after 1 and 21 d retention for Sharkey soil. Isotherms are also presented in the presence of 100 mg of P in the soil solution.

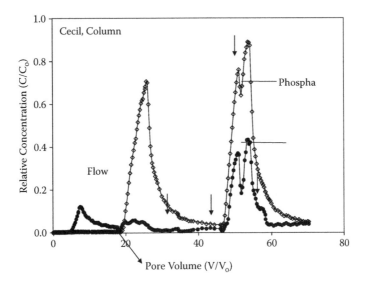

FIGURE 7.37
Breakthrough curve results of V and P from Cecil soil (column 1). The pulse sequence was V, P, and V and P sequentially.

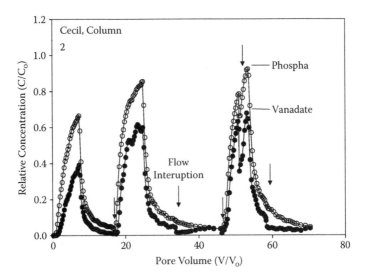

FIGURE 7.38
Breakthrough curve results of V and P from Cecil soil (column 2). The pulse sequence was P, V, and V and P sequentially.

arrival time for each BTC, Cecil soil exhibited higher affinity for V than P. This result was consistent for all three consecutive pulses. The location of the BTC maximum was the same for both P and V. This suggests that the retention mechanisms for P as well as V on Cecil soil were similar. However, a major difference is that the amount that appears to be irreversibly sorbed (or slowly released) is larger for V than P. We infer the extent of irreversible sorption (or slow release) based on the area under the BTCs. Moreover, the extensive tailing during leaching or desorption also suggests nonlinear and/ or kinetic retention mechanisms for both V and P (Mansell et al., 1992).

Competition between V and P for the same retention sites on soil surfaces is evident when one compares the BTCs shown in Figure 7.37 with those in Figure 7.38. The absence of P in the pulse solution resulted in a much higher retention of V, with a much lower peak concentration when compared to the case when both P and V were present in the input pulse solution. The extent of retention can be estimated from the area under the BTC for V in the first input pulse where four times the amount of V was found in the effluent solution when P was present. In addition, a subsequent P pulse displaced a significant portion of V already retained by the Cecil soil during the first pulse. However, the opposite was not as apparent when a P pulse followed a V pulse, as shown in Figure 7.39.

Phosphate-soil interactions exhibit various reactions, including kinetic, reversible, and nonlinear sorption, often with some degree of irreversibility (Mansell et al., 1992; Chen et al., 1996). Nonlinear sorption occurs primarily due to reactions with Fe and Al oxides in the soil. The term sorption is used in a general sense here to include P adsorption by mineral surfaces

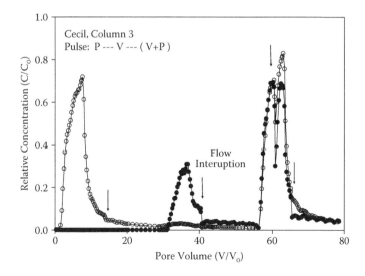

FIGURE 7.39
Breakthrough curve results of V and P from Cecil soil (column 3). The column received three pulses of a mixed solution of P and V.

and complexation with humic acids (Al and Fe may form bridges between organic ligands and P ions). Irreversible sorption is considered to involve chemisorption and fixation within mineral structures. Scientists generally agree that many types of P retention processes can take place simultaneously. Less is known as to whether these reactions operate in series or in parallel. Experimental techniques are available in the literature to monitor sorption and fixation reactions. Although a mathematical model to completely describe P reactions in soil doesn't exist, models that approximate these reactions do. Multireaction sorption models constitute a class of such approximations and offer a practical means of approximating P mobility during water flow in soils. In general, the sorption behavior of vanadate in soils is similar to inorganic phosphate. However, V also reduces acid phosphatase activity in soil, which alters the rate of mineralization of organic matter and may reduce phosphate bioavailability (Tyler, 1976). The interactions of V with other ions present in the soil solution (e.g., phosphate) and its potential mobility in the soil profile have not been fully investigated (Mikkonen and Tummavuori, 1994). For all three columns with an input pulse containing both P and V BTCs for P arrived earlier and with higher peak concentrations than those for V. In addition, a decrease in effluent concentration due to flow interruption was observed in all BTCs associated with the third pulse in all three columns, which indicates increases in the amounts of P or V sorbed (Figures 7.37 to 7.39). These observations are consistent for both P and V and show the kinetic nature of the retention mechanisms. For Sharkey clay soil, miscible displacement results illustrated the extent of P as well as V retention in this montmorillonitic soil (results are not shown). For all input V concentrations, the presence of P resulted in a decrease in the amount of V sorbed at all reaction times. This finding was consistent with that obtained for the kaolinitic Cecil soil as well as a montmorillonitic Sharkey soil.

7.8.5 Copper–Magnesium

Copper is a heavy metal exploited in large quantities for economic value which often leaves many abandoned Cu mines around the world. Drainage water from the abandoned mines often contains high Cu concentrations and can be of environmental concern (Amacher et al., 1995). An understanding of the retention and transport of Cu in mine soils is necessary for minimizing possible adverse effects from Cu mining.

We present retention and transport studies for two soils: Cecil soil and McLaren soil (with pH of 4.1, organic matter content of 3.03%, and CEC of 3.3 meq/100 g). The Cecil soil was chosen as a benchmark soil and was previously characterized for its affinity for retention of several heavy metals (Buchter et al., 1989). McLaren soil was obtained from a site near an abandoned Cu mine on Fisher Mountain, Montana. Acid mine drainage from the abandoned mine flows into Daisy Creek below the mine, which is located about two miles from Cooke City, Montana (Amacher et al. 1995). Results

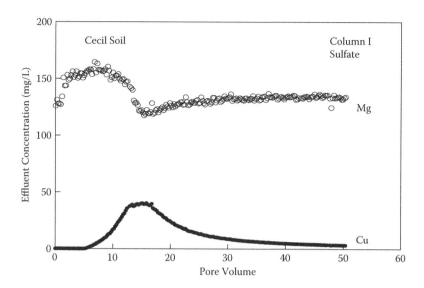

FIGURE 7.40
Breakthrough curves for Cecil soil (column I) with sulfate as the counterion.

from undisturbed soils near the minespoil are reported here. The minespoil is characterized by high Cu contents and low pH.

Figures 7.40 and 7.41 show the effect of total concentration or ionic strength of the input pulse solution on Cu breakthrough results. When Cu was introduced in Mg background solution with minimum change in ionic strength,

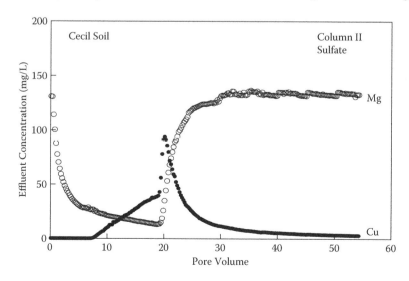

FIGURE 7.41
Breakthrough curves for Cecil soil (column II) with sulfate as the counterion.

Cu BTCs appear symmetrical in shape with considerable tailing and a peak concentration of 40 mg L⁻¹. The Mg BTC shows an initial increase in concentration due to slight increase in ionic strength followed by a continued decrease during leaching. When Cu was introduced in the absence of a background solution, the total concentration decreased from 0.005 to 0.0015 M. As shown in Figure 7.41, the Cu BTC showed a sharp increase in concentration due to the chromatographic (or snow plow) effect (Selim et al., 1992). The peak Cu concentration was 94 mg L⁻¹, and the corresponding Mg concentration in the effluent decreased due to depletion of Mg during the introduction of Cu. The Mg concentration increased thereafter to a steady-state level during subsequent leaching. This snowplow effect is a strong indication of competitive ion exchange between Mg and Cu cations. The amount of Cu recovered in the effluent was 53% of that applied in the presence of MgSO₄ as the background solution, whereas only 38% was recovered when no background solution was used. Therefore, miscible displacement experiments indicated that there was strong ion exchange between Cu and Mg cations, which was also affected by the counterion used. Effluent peak concentrations were three- to fivefold that of the input Cu pulse, which is indicative of pronounced chromatographic effect.

7.8.6 Cadmium–Phosphate

The influence of the presence of P on Cd sorption in soils is illustrated by the family of Cd isotherms shown in Figure 7.42 where different initial P

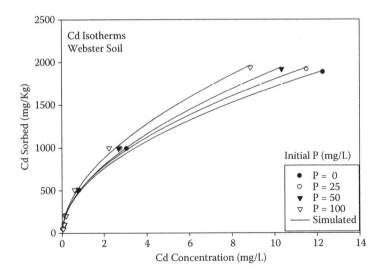

FIGURE 7.42
Cadmium adsorption isotherms for Webster soil after 24 h of sorption in the presence of various P concentrations. Solid curves are simulations using the Freundlich equation.

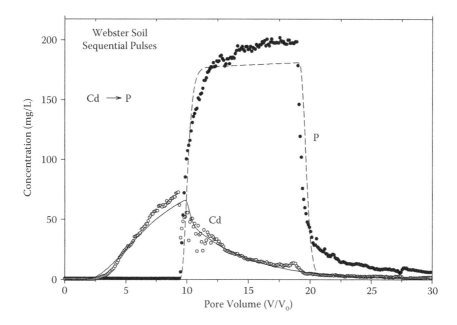

FIGURE 7.43
Cadmium and P breakthrough curve results for a Webster soil column that received pulses of Cd and P. The solid and dashed curves are multireaction model predictions.

concentrations were added (0 to 100 mg/L). All Cd isotherms appear non-linear regardless of the initial P concentration in solution in Webster soil. The influence of P on increased Cd sorption was clearly manifested by the increased Freundlich distribution coefficient K_F from 532 to 601 ml/g for initial P concentrations of 0 and 100 ml/g. In contrast, the nonlinear parameter N did not exhibit much variation (0.50 to 0.54) for all P concentrations.

In Figure 7.43, the BTCs for Cd and P are presented for a Webster soil column that received a Cd pulse followed by a P pulse, that is, sequential pulses. Figure 7.44 shows BTCs for Cd and P for a Webster soil column that received two consecutive pulses of a mixed P and Cd solution. Cadmium BTCs indicate strong retention with low Cd concentration in the effluent solution, especially for the mixed column. Peak concentration of the BTC results was 70 mg/L (C/C₀ of 0.37); compared to 30 and 60 mg/L associated with the first and second pulses of the mixed column. Based on the area under curve, the amount of Cd recovery was 28% for the sequential column and only 18% for the mixed column, which indicates higher sorption of Cd in the presence of P. The observed strong Cd retention exhibited by the results of Figures 7.43 and 7.44 is supported by the sorption isotherm shown in Figure 7.42, which indicates high Cd sorption, which is likely due to high clay content, organic matter, amorphous Fe and Al, and the presence of carbonates. In addition,

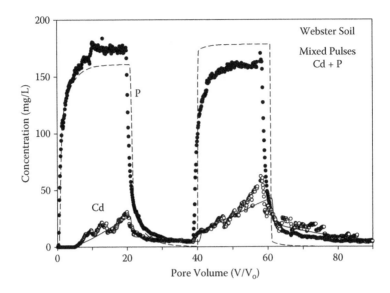

FIGURE 7.44

Cadmium and P breakthrough curve results for a Webster soil column that received two mixed pulses of Cd and P solution. The solid and dashed curves are multireaction model predictions.

x-ray diffraction analysis indicates that for Webster soil, smectite is the dominant clay type.

The solid curves shown in Figures 7.43 and 7.44 are multireaction transport model (MRTM) simulations for the two Webster columns. For the sequential column, the model provided a good description of Cd as well as P results, with R^2 of 0.96 for Cd and 0.95 for P. For the mixed column, the respective values were 0.85 and 0.86. Overall good model description of the BTCs for Cd and P were achieved for the sequential and mixed columns. Peak P concentration and the tailing of the BTC were not well described by the model, however. The model version that provided the best prediction for Cd was that which assumed only kinetic reactions. Other model versions that account for equilibrium sorption resulted in the finite-difference solution of the convection-dispersion transport equation being unstable and convergence was not attained. This is an indication that all Cd retention mechanisms for Webster soil are dominated by kinetic or time-dependent reactions. For P, the dashed curves shown in Figures 7.43 and 7.44, the MRTM versions were fully kinetic which included reversible and irreversible P retention. The breakthrough results for P indicate that for both the sequential and mixed columns more than 90% P recovery was obtained. These results were not expected and a significant amount of P sorption was anticipated based on the soil properties of this soil. In fact the lack of P sorption by the soil columns is in contrast to strong sorption of P for this soil with K_F of 134 and b of 0.682. These results suggest that the presence of Cd resulted in reduced P retention and enhanced mobility and bioavailability in this soil. Similar findings were obtained for three other soils having a wide range of soil properties.

7.8.7 Silver–Zinc

Silver ion is one of the most toxic forms of all heavy metals, surpassed only by mercury, and thus has been assigned to the highest toxicity class, together with cadmium, chromium, copper, and mercury (Ratte, 1998). The principle sources of silver are zinc, copper, gold, and lead mines. Because of its wide use, silver is regarded as a significant metal and effort over the last several decades has been devoted to understanding the geochemical reactions of silver (Ag) with other heavy metals in the soil-water environment. Because of increased use of silver in different industries, including the pharmaceutical industry, discharges containing silver are often reported. Release from industry and from disposal of sewage sludge and refuse are major sources of soil contamination with silver. The major source of elevated silver levels in cultivated soils is from the application of sewage sludge and sludge effluents as agricultural amendments.

Examples illustrating the extent of mobility of Ag and the influence of the presence of Zn on Ag mobility in Windsor soil are presented in Figures 7.45 to 7.47. Ag and Zn were applied to the soil column as $Ag(NO_3)_2$ and $Zn(NO_3)_2$. The BTC for Ag illustrates extensive transport of Ag with high peak concentrations regardless of the presence of Zn. In Figure 7.46, BTCs for Ag and Zn are presented from the mixed pulse and show early arrival in the effluent

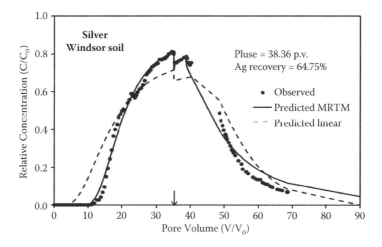

FIGURE 7.45
Breakthrough curve results for Ag from a Windsor soil column that received a pulse of Ag solution. Solid and dashed curves are simulations using the multireaction and linear models.

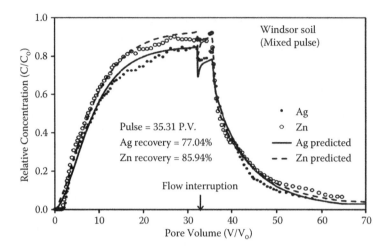

FIGURE 7.46
Breakthrough curve results for Ag and Zn from a Windsor soil column that received a mixed pulse with Ag and Zn solution. Solid curves are simulations using the multireaction model.

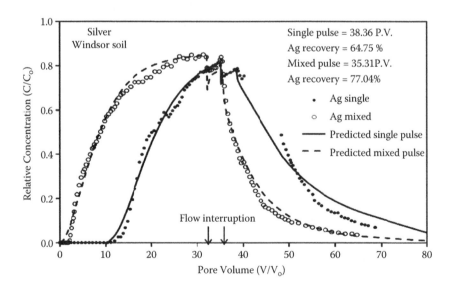

FIGURE 7.47
Comparison of breakthrough curves for Ag from two Windsor soil columns. One received a Ag pulse only; the second received a mixed pulse of Ag and Zn solution. Solid and dashed curves are simulations using a multireaction model.

solution (1.8 pore volumes) for both Zn and Ag. In Figure 7.47, significant difference in the arrival time of Ag in the presence and absence of Zn is illustrated. Moreover, the presence of Zn resulted in a slight increase of Ag sorption with an influent recovery of 35% compared to 38% in the absence of Zn.

The solid and dashed curves shown in Figures 7.45 to 7.47 are multireaction model simulations of Ag and Zn in the Windsor soil columns. Good model predictions for both Ag and Zn were obtained as depicted by the prediction of the effluent front, peak concentration maxima, and the slow release during leaching. However, when a linear model was used, predictions were considered less than adequate. In addition, a decrease in effluent concentration due to flow interruptions was observed in all BTCs. These are consistent for both Ag and Zn and show the kinetic nature of their retention mechanisms, which were well predicted by the multireaction models.

References

Abd-Elfattah, A., and K. Wada. 1981. Adsorption of lead, copper, zinc, cobalt, and calcium by soil that differ in cation-exchange materials. *J. Soil Sci.* 32: 271–283.

Adriano, D. C. 2001. *Trace Elements in Terrestrial Environments: Biogeochemistry, Bioavailability and Risk of Metals*, second edition. New York: Springer.

Agbenin, J. O. 1998. Phosphate-induced zinc retention in a tropical semi-acid soil. *European J. Soil Sci.* 49: 693–700.

Ahumada, I. T., A. Bustamante, and E. B. Schalscha. 1997. Zinc speciation in phosphate-affected soils. *Commun. Soil Sci. Plant Anal.* 28: 989–995.

Allison, J. D., D. S. Brown, and K. J. Novo-Gradac. 1991. MINTEQA2/PRODEFA2, A Geochemical Assessment Model for Environmental Systems: Version 3.0 User's Manual. EPA/600/3-91/021. U.S. Environmental Protection Agency, Washington, DC.

Amacher, M. C., R. W. Brown, R. W. Sidle, and J. Kotuby-Amacher. 1995. Effect of mine waste on element speciation in headwater streams. In *Metal Speciation and Contamination of Soil,* edited by H. E. Allen et al., 275–309. Boca Raton, FL: Lewis Publishers.

Amacher, M. C., H. M. Selim, and I. K. Iskandar. 1988. Kinetics of chromium(VI) and cadmium retention in soils: A nonlinear multireaction model. *Soil Sci. Soc. Am. J.,* 52: 398–408.

Anke, M. 2004. Vanadium. In *Elements and Their Compounds in the Environment*, second edition, edited by E. Merian, M. Anke, M. Ihnat, and M. Stoeppler, 1171–1191. Weinheim: Wiley.

Antoniadis, V., and C. D. Tsadilas. 2007. Sorption of cadmium, nickel and zinc in mono- and multimetal systems. *Appl. Geochem.* 22: 2375–2380.

Atanassova, I. 1999. Competitive effect of copper, zinc, cadmium and nickel on ion adsorption and desorption by soil clays. *Water Air. Soil Poll.* 113: 115–125.

Barnett, M. O., P. M. Jardine, S. C. Brooks, and H. M. Selim. 2000. Adsorption and transport of uranium(VI) in subsurface media. *Soil Sci. Soc. Am. J.* 64: 908–917.

Barrow, N. J. 1987. The effect of phosphate on zinc sorption by soils. *J. Soil Sci.* 38: 453–459.

Barrow, N. J., J. Gerth, and G. W. Brummer. 1989. Reaction-kinetics of the adsorption and desorption of nickel, zinc and cadmium by goethite. II. Modeling the extent and rate of reaction. *J. Soil Sci.* 40: 437–450.

Barrow, N. J. 1992. The effect of time on the competition between anions for sorption. *J. Soil Sci.* 43: 421–428.

Barrow, N. J., P. Cartes, and M. L. Mora. 2005. Modifications to the Freundlich equation to describe anion sorption over a large range and to describe competition between pairs of ions. *Eur. J. Soil Sci.* 56: 601–606.

Bibak, A. 1997. Competitive sorption of copper, nickel, and zinc by an oxisol. *Commun. Soil Sci. Plant Anal.* 28: 927–937.

Bittel, J. E., and R. J. Miller. 1974. Lead, cadmium, and calcium selectivity coefficients of a montmorillonite, illite, and kaolinite. *J. Environ. Qual.* 3: 250–253.

Bond, W. J., and I. R. Phillips. 1990. Approximate solution for cation transport during unsteady, unsaturated soil water flow. *Water Resour. Res.* 26: 2195–2205.

Bolland, M. D. A., A. M. Posner, and J. P. Quirk. 1977. Zinc adsorption by goethite in the absence or presence of phosphate. *Australian J. Soil Res.* 15: 279–286.

Buchter, B., B. Y. Davidoff, M. C. Amacher, C. Hinz, I. K. Iskandar, and H. M. Selim. 1989. Correlation of Freundlich Kd and *n* retention parameters with soils and elements. *Soil Sci.* 148: 370–379.

Cavallaro, N., and M. B. McBride. 1984. Zinc and copper sorption and fixation by an acid soil clay: Effect of selective dissolutions. *Soil Sci. Soc. Am. J.* 48: 1050–1054.

Chen, J. S., R. S. Mansell, P. Nkedi-Kizza, and B. A. Burgoa. 1996. Phosphorus transport during transient, unsaturated water flow in an acid sandy soil. *Soil Sci. Soc. Am. J.* 60: 42–48.

Coats, K. H., and B. D. Smith. 1964. Dead-end pore volume and dispersion in porous media. *Soc. Pet. Eng.* 4: 73–84.

Crans, D. C., J. J. Smee, E. Gaidamauskas, and L. Yang. 2004. The chemistry and biochemistry of vanadium and the biological activities exerted by vanadium compounds. *Chem. Rev.* 104: 849–902.

Darland, J. E., and W. P. Inskeep. 1997. Effects of pH and phosphate competition on the transport of arsenate. *J. Environ. Qual.* 26: 1133–1139.

De Brouwere, K., E. Smolders, and R. Merckx. 2004. Soil properties affecting solid-liquid distribution of As(V) in soils. *Eur. J. Soil Sci.* 55: 165–173.

Dechaine, G. P., and M. P. Gray. 2010. Chemistry and association of vanadium compounds in heavy oil and bitumen, and implications for their selective removal. *Energy Fuels* 24: 2795–2808.

Echeverría, J. C., M. T. Morera, C. Mazkiarán, and J. J. Garrido. 1998. Competitive sorption of heavy metal by soils. Isotherms and fractional factorial experiments. *Environ. Pollut.* 101: 275–284.

Eick, M. J., B. R. Naprstek, and P. V. Brady. 2001. Kinetics of Ni(II) sorption and desorption on Kaolinite: Residence time effects. *Soil Sci.* 166: 11–17.

Elbana, T. A., and H. M. Selim. 2011. Copper mobility in acidic and alkaline soils: Miscible displacement experiments. *Soil Sci. Soc. Am. J.* 75: 2101–2110.

Ford, R. G., A. C. Scheinost, K. G. Scheckel, and D. Sparks. 1999. The link between clay mineral weathering and the stabilization of Ni surface precipitates. *Environ. Sci. Technol.* 33: 3140–3144.

Fuller, C. C., J. A. Davis, and G. A. Waychunas. 1993. Surface chemistry of ferrihydrite: Part 2. Kinetics of arsenate adsorption and coprecipitation. *Geochim. Cosmochim. Acta* 57: 2271–2282.

Gaines, G. L., and H. C. Thomas. 1953. Adsorption studies on clay minerals. II. A formulation of the thermodynamics of exchange adsorption. *J. Chem. Phys.* 21: 714–718.

Gaston, L. A., and H. M. Selim. 1990a. Transport of exchangeable cations in an aggregated clay soil. *Soil Sci. Soc. Am. J.* 54: 31–38.

Gaston, L. A., and H. M. Selim. 1990b. Prediction of cation mobility in montmorilonitic media based on exchange selectivities of montmorillonite. *Soil Sci. Soc. Am. J.* 54: 1525–1530.

Gaston, L. A., and H. M. Selim. 1991. Predicting cation mobility in kaolinite media based on exchange selectivities of kaolinite. *Soil Sci. Soc. Am. J.* 55: 1255–1261.

Gaudalix, M. E.. and M. T. Pardo. 1995. Zinc sorption by acid tropical soils as affected by cultivation. *J. Soil Sci.* 46: 317–322.

Goldberg, S., and L. J. Criscenti. 2008. Modeling adsorption of metals and metalloids by soil components. In Biophysico-chemical *Processes of Heavy Metals and Metalloids in Soil Environments*, edited by A. Violante, P. M. Huang, and G. M. Gadd. New York: John Wiley & Sons.

Gomes, P. C., M. P. F. Fontes, and A. G. da Silva. 2001. Selectivity sequence and competitive adsorption of heavy metals by Brazilian soils. *Soil Sci. Soc. Am. J.* 65: 1115–1121.

Grossl, P. R., M. J. Eick, D. L. Sparks, S. Goldberg, and C. C. Ainsworth. 1997. Arsenate and chromate retention mechanisms on goethite. II. Kinetic evaluation using a pressure-jump relaxation technique. *Environ. Sci. Technol.* 31: 321–326.

Gutierrez, M., and H. R. Fuentes. 1993. Modeling adsorption in multicomponent systems using a Freundlich-type isotherm. *J. Contam. Hydrol.* 14: 247–260.

Harmsen, K. 1977. *Behavior of Heavy Metals in Soils*. Wageningen, The Netherlands: Centre for Agriculture Publishing and Documentation.

Harter, R. S., and R. Naidu. 2001. An assessment of environmental and solution parameter impact on trace-metal sorption by soils. *Soil Sci. Soc. Am. J.* 65: 597–612.

Hingston, F. J., A. M. Posner, and J. P. Quirk. 1971. Competitive adsorption of negatively charged ligands on oxide surfaces. *Disc. Faraday Soc.* 52: 334–342.

Hinz, C., and H. M. Selim. 1994. Transport of zinc and cadmium in soils: Experimental evidence and modeling approaches. *Soil Sci. Soc. Am. J.* 58: 1316–1327.

Jain, A., and R. H. Loeppert. 2000. Effect of competing anions on the adsorption of arsenate and arsenite by ferrihydrite. *J. Environ. Qual.* 29: 1422–1430.

Jardine, P. M., and D. L. Sparks. 1984. Potassium-calcium exchange in a multireactive soil system. I. Kinetics. *Soil Sci. Soc. Am. J.* 48: 39–45.

Jeon, B., B. A. Dempsey, and W. D. Burgos. 2003. Sorption kinetics of Fe(II), Zn(II), Co(II), Ni(II), Cd(II), and Fe(II)/Me(II) onto hematite. *Water Res.* 37: 4135–4142.

Jia, Y. F., L. Y. Xu, Z. Fang, and G. P. Demopoulos. 2006. Observation of surface precipitation of arsenate on ferrihydrite. *Environ. Sci. Technol.* 40: 3248–3253.

Kretzschmar, R., and A. Voegelin. 2001. Modeling competitive sorption and release of heavy metals in soils. In *Heavy Metals Release in Soils*, edited by H. M. Selim, and D. L. Sparks, 55–88. Boca Raton, FL: Lewis Publishers.

Kuo, S., and B. L. McNeal.1984. Effect of pH and phosphate on cadmium sorption by hydrous ferric oxide. *Soil Sci. Soc. Am. J.* 48: 1040–1044.

Kurdi, E., and H. E. Doner. 1983. Zinc and copper sorption and interaction in soils. *Soil Sci. Soc. Am. J.* 47: 873—876.

Liao, L., and H. M. Selim. 2010. Reactivity of nickel in soils: Evidence of retention kinetics. *J. Environ. Qual.* 39: 1290–1297.

Liu, C., T. Chang, M. Wang, and C. Huang. 2006. Transport of cadmium, nickel, and zinc in Taoyuan red soil using one-dimensional convective-dispersive model. *Geoderma* 131: 181–189.

Liu, F., A. De Cristofaro, and A. Violante. 2001. Effect of pH, phosphate and oxalate on the adsorption/desorption of arsenate on/from goethite. *Soil Sci.* 166: 197–208.

Manning, B. A., and S. Goldberg. 1996a. Modeling competitive adsorption of arsenate with phosphate and molybdate on oxide minerals. *Soil Sci. Soc. Am. J.* 60: 121–131.

Manning, B. A., and S. Goldberg. 1996b. Modeling arsenate competitive adsorption on kaolinite, montmorillonite and illite. *Clays Clay Miner.* 44: 609–623.

Mansell, R. S., S. A. Bloom, H. M. Selim, and R. D. Rhue. 1988. Simulated transport of multiple cations in soil using variable selectivity coefficients. *Soil Sci. Soc. Am. J.* 52: 1533–1540.

Mansell, R. S., S. A. Bloom, B.A. Burgoa, P. Nkedi-Krzza, and J. S. Chen. 1992. Experimental and simulated P transport in soil using a multireaction model. *Soil Sci.* 153: 185–194.

Mesquita, M. E., and J. M. Vieira e Silva. 2002. Preliminary study of pH effect in the application of Langmuir and Freundlich isotherms to Cu–Zn competitive adsorption. *Geoderma* 106: 219–234.

Mikkonen, A., and J. Tummavuori. 1994. Retention of vanadium (V) by three Finnish mineral soils. *Eur. J. Soil Sci.* 45: 361–368.

Murali, V., and L. A. G. Aylmore. 1983. Competitive adsorption during solute transport in soils. III. A review of experimental evidence of competitive adsorption and an evaluation of simple competition models. *Soil Sci.* 136: 279–290.

Parker, J. C., and P. M. Jardine. 1986. Effect of heterogeneous adsorption behavior on ion transport. *Water Resourc. Res.* 22: 1334–1340.

Parkhurst, D. L., and C. A. J. Appelo. 1999. User's guide to PHREEQC Version 2: A computer program for speciation, batch reaction, one-dimensional transport and inverse geochemical calculations. In *Water Resources Investigations*, Report 99–4259.Lakewood, CO: U.S. Geological Survey.

Papini, M. P., T. Saurini, and A. Bianchi. 2004. Modeling the competitive adsorption of Pb, Cu, Cd and Ni onto a natural heterogeneous sorbent material (Italian "Red Soil"). *Ind. Eng. Chem. Res.* 43: 5032–5041.

Pardo, M. T. 1999. Influence of phosphate on zinc reaction in variable charge soils. *Commun. Soil Sci. Plant Anal.* 30: 725–737.

Peryea, F. J. 1991. Phosphate-induced release of arsenic from soils contaminated with lead arsenate. *Soil Sci. Soc. Am. J.* 55: 1301–1306.

Pigna, M., G. S. R. Krishnamurti, and A. Violante. 2006. Kinetics of arsenate sorption-desorption from metal oxides: Effect of residence time. *Soil Sci. Soc. Am. J.* 70: 2017–2027.

Puls, R. W., and H. L. Bohn. 1988. Sorption of cadmium, nickel, and zinc by kaolinite and montmorillonite suspensions. *Soil Sci. Soc. Am. J.* 52: 1289–1292.

Ratte, H., 1998. Bioaccumulation and toxicity of silver compounds: A review. *Environ. Toxic. Chem.*18: 89–108.

Raven, K. P., A. Jain, and R. H. Loeppert. 1998. Arsenite and arsenate adsorption on ferrihydrite: Kinetics, equilibrium, and adsorption envelopes. *Environ. Sci. Technol.* 32: 344–349.

Roy, W. R., J. J. Hassett, and R. A. Griffin. 1986a. Competitive coefficient for the adsorption of arsenate, molybdate, and phosphate mixture by soils. *Soil Sci. Soc. Am. J.* 50: 1176–1182.

Roy, W. R., J. J. Hassett, and R. A. Griffin. 1986b. Competitive interactions of phosphate and molybdate on arsenate adsorption. *Soil Sci.* 142: 203–210.

Rupa, T. R., and K. P. Tomar. 1999. Zinc desorption kinetics as influenced by pH and phosphorus in soils. *Commun. Soil Sci. Plant Anal.* 30(13&14): 1951–1962.

Saeed, M., and R. L. Fox. 1979. Influence of phosphate fertilization on zinc adsorption by tropical soils. *Soil Sci. Soc. Am. J.* 43: 683–686.

Sarret, G., P. Saumitou-Laprade, V. Bert, O. Proux, J. Hazemann, A. Traverse, M. A. Marcus, and A. Manceau. 2002. Forms of zinc accumulated in the hyperaccumulator *Arabidopsis*. *Plant Physiol.* 130: 1815–1826.

Scheidegger, A. M., G. M. Lamble, and D. L. Sparks. 1996. Investigation of Ni sorption on pyrophyllite: An XAFS study. *Environ. Sci. Technol.* 30: 548–554.

Schulthess, C. P., and C. P. Huang. 1990. Adsorption of heavy-metals by silicon and aluminum-oxide surfaces on clay-minerals. *Soil Sci. Soc. Am. J.* 54: 679–688.

Selim, H. M. 1992. Modeling the transport and retention of inorganics in soils. *Adv. Agron.* 47: 331–384.

Selim, H. M., and M. C. Amacher. 1997. *Reactivity and Transport of Heavy Metals in Soils.* Boca Raton, FL: CRC Press.

Selim, H. M., B. Buchter, C. Hinz, and L. Ma. 1992. Modeling the transport and retention of cadmium in soils: Multireaction and multicomponent approaches. *Soil Sci. Soc. Am. J.* 56: 1004–1015.

Serrano, S., F. Garrido, C. G. Campbell, and M. T. Garcia-Gonzalez. 2005. Competitive sorption of cadmium and lead in acid soils of Central Spain. *Geoderma* 124: 91–104.

Sheindorf, C., M. Rebhun, and M. Sheintuch. 1981. A Freundlich-type multicomponent isotherm. *J. Colloid Interface Sci.* 79: 136–142.

Smedley, P. L., and F. G. Kinniburgh. 2002. A review of the source, behaviour and distribution of arsenic in natural waters. *Appl. Geochem.* 17: 517–568.

Smith, E., R. Naidu, and A. M. Alston. 2002. Chemistry of inorganic arsenic in soils. II. Effect of phosphorous, sodium, and calcium on arsenic sorption. *J. Environ. Qual.* 31: 557–563.

Smith, S. L., and P. R. Jaffé. 1998. Modeling the transport and reaction of trace metals in water-saturated soils and sediments. *Water Resour. Res.* 34: 3135–3147.

Sparks, D. L. 1989. *Kinetics of Soil Chemical Processes.* San Diego, CA: Academic Press.

Sparks, D. L. 1998. *Soil Physical Chemistry*, second edition. Boca Raton, FL: CRC Press.

Sparks, D. L. 2003. *Environmental Soil Chemistry*, second edition. San Diego, CA: Academic Press.

Sposito, G. 1984. *The Surface Chemistry of Soils.* New York: Oxford University Press.

Tagwira, F., M. Piha, and L. Mugwira. 1993. Zinc studies in Zimbabwean soils: Effect of pH and phosphorus on zinc adsorption by two Zimbabwean soils. *Commun. Soil Sci. Plant Anal.* 24: 701–716.

Toride , N., F. J. Leij, and M. Th. van Genuchten. 1995. The CXTFIT code for estimating transport parameters from laboratory or field tracer experiments, Version 2.0, Research Report No. 137, U.S. Salinity Laboratory, USDA-ARS, Riverside, CA.

Tyler, G. 1976. Influence of vanadium on soil phosphatase activity. *J. Environ. Qual.* 5: 216–217.

Underwood, E. J. 1977. *Trace Elements in Human and Animal Nutrition.* New York: Academic Press.

Violante, A., and M. Pigna. 2002. Competitive sorption of arsenate and phosphate on different clay minerals and soils. *Soil Sci. Soc. Am. J.* 66: 1788–1796.

Voegelin, A., and R. Kretzschmar. 2005. Formation and dissolution of single and mixed Zn and Ni precipitates in soil: Evidence from column experiments and extended X-ray absorption fine structure spectroscopy. *Environ. Sci. Technol.* 39: 5311–5318.

Waltham, C. A., and W. J. Eick. 2002. Kinetics of arsenic adsorption on goethite in the presence of sorbed silicic acid. *Soil Sci. Soc. Am. J.* 66: 818–825.

Wang, J. J., and D. L. Harrell. 2005. Effect of ammonium, potassium, and sodium cations and phosphate, nitrate, and chloride anions on zinc sorption and lability in selected acid and calcareous soils. *Soil Sci. Soc. Am. J.* 69: 1036–1046.

Wilson, G. V., F. E. Rhoton, and H. M. Selim. 2004. Modeling the impact of ferrihydrite on adsorption-desorption of soil phosphorus. *Soil Sci.* 169: 271–281.

Woolson, E. A., J. H. Axley, and P. C. Kearney. 1973. The chemistry and phytotoxicity of arsenic in soils: Effect of time and phosphorous. *Soil Sci. Soc. Am. Proc.* 37: 254–259.

Wu, C.-H., C.-Y. Kuo, C.-F. Lin, and S.-L. Lo. 2002. Modeling competitive adsorption of molybdate, sulfate, selenate, and selenite using a Freundlich-type multi-component isotherm. *Chemosphere* 47: 283–292.

Xie, R. J., and A. F. Mackenzie. 1989. Effects of sorbed orthophosphate on zinc status in three soils of eastern Canada. *J. Soil Sci.* 40: 49–58.

Zhang, H., and H. M. Selim. 2007. Modeling competitive arsenate-phosphate retention and transport in soils: A multi-component multi-reaction approach. *Soil Sci. Soc. Am. J.* 71: 1267–1277.

Zhang, H., and H. M. Selim. 2008. Competitive sorption-desorption kinetics of arsenate and phosphate in soils. *Soil Sci.* 173: 3–12.

Zhao, H., and R. Stanforth. 2001. Competitive adsorption of phosphate and arsenate on goethite. *Environ. Sci. Technol.* 35: 4753–4757.

Zhao, K., and H. M. Selim. 2010. Adsorption-desorption kinetics of Zn in soils: Influence of phosphate. *Soil Sci.* 75: 145–153.

8

Mobile, Immobile, and Multiflow Domain Approaches

Nonequilibrium transport of solute in soils and geological media is commonly attributed to two main processes: (1) the rate-limited mass transfer of aqueous species to or from the regions with limited or no advective flow; and (2) the time-dependent chemical reaction at the surfaces of solid materials. The first process is often referred to as transport-related nonequilibrium or physical nonequilibrium and the second process is referred to as reaction-related nonequilibrium or chemical nonequilibrium. As a result of pedogenic processes, soils and geological media are often characterized by extensive chemical and physical heterogeneities. Soil heterogeneity has a profound effect on the transport of chemicals in the soil profile. The physical heterogeneity is mainly attributed to the highly nonuniform pore space as a result of the soil aggregates, soil layers, as well as cavities developed from natural and anthropogenic activities. The highly irregular flow field resulting from the spatial variance of pore sizes and connectivity is a major driving force of the nonequilibrium transport of solutes in soils (Biggar and Nielsen, 1976; Dagan, 1984). Conceptually, the nonequilibrium transport of nonreactive solutes in structured soils can be described as a rapid movement of water flow and solutes within preferential flow paths and a diffusive mass transfer of solutes to a stagnant region. This mobile-immobile dual-region concept is the foundation of the mathematical models used for simulating the physical nonequilibrium transport (van Genuchten, 1981; Reedy et al., 1996; Simunek et al., 2003).

8.1 Matrix Diffusion

In general, the nonequilibrium mass transfer of solutes in soils proceeds in two stages: (1) external or film diffusion from flow water to the solid interface; and (2) internal diffusion through the porous network of soil aggregates (Brusseau, 1993). Film diffusion is of importance only at low flow velocities. Under environmental conditions, diffusive mass transfer in the micropores inside soil aggregates is an important process influencing the mobility of reactive metals in soils. The intra-aggregate diffusion serves as a reservoir in the transport of contaminants by spreading them from flowing water to the immobile region of stagnant pore water. Furthermore, the diffusion of reactive solutes inside soil matrix increases their contact with reaction sites on the surfaces of minerals and organic matters. The time-dependent process can significantly impact the extent and rate of geochemical and microbial reactions with the solid phase.

Matrix diffusion can be manifested in the laboratory by examining the asymmetric breakthrough curves from miscible displacement experiments with nonreactive tracers. Experimental and modeling studies have demonstrated that the mass transfer process can be influenced by aggregate shape, particle size distribution, and pore geometries. Brusseau (1993) performed column experiments using four porous media with different physical properties to investigate the influence of solute size, pore water velocity, and intraparticle porosity on solute dispersion and transport in soil. It was concluded that solute dispersion in aggregated soil was caused by hydrodynamic dispersion, film diffusion, intraparticle diffusion, and axial diffusion at low pore water velocities, whereas for sandy soils, the contribution of intraparticle diffusion and film diffusion is negligible and hydrodynamic dispersion is the predominant source of dispersion for most conditions.

The flow interruption technique has been employed to detect and quantify the rate and extent of physical nonequilibrium during solute transport in heterogeneous porous media. A flow interruption stops the injection of fluids and tracers, allowing more time for reactive tracers to interact with the solid phase and for all tracers to diffuse into or out of the matrix. Reedy et al. (1996) conducted nonreactive tracer (Br⁻) miscible displacement experiments with a large, undisturbed soil column of weathered, fractured shale from a proposed waste site on the Oak Ridge Reservation. The tracer flow was interrupted for a designated time to quantify the diffusive mass transfer of a nonreactive solute between the matrix porosity and preferential flow paths in fractured subsurface media. Decreased and increased tracer concentrations were observed after flow interruption during tracer infusion and displacement, respectively, when flow was reinitiated.

8.2 Preferential Flow

Preferential flow is primarily due to the naturally occurring soil structure, which consists of pores having different diameters, cracks formed by soil shrinking during drying and wetting cycles, various macropores, and channels created by decaying plant roots and earthworms. Luxmoore (1981) suggested that macropores are those with diameters greater than 0.1 mm in size. In shrink-swell clay soils such as Sharkey clay, cracks 2 to 4 cm in width and 12 to 15 cm deep are not uncommon during drying periods. Preferential or macropore flow is of particular significance in heavy textured soils because they can increase infiltration and may result in bypass flow where water and solutes move rapidly through the soil profile with limited interaction with the soil matrix.

Macropores are essentially considered as noncapillary pores, and it can be assumed that flow occurs in macropores only in the presence of free water at the soil surface. However, several observations indicate that macropore flow occurs in very dry soils at the onset of a rain. This is of significance in the movement of reactive chemicals to lower depth due to water flow in macropores. In general, preferential flow occurring via distinct flow pathways in fractures and macropores can significantly influence the movement of heavy metals through soils. Preferential and nonequilibrium flow and transport are often difficult to characterize and quantify because of the highly irregular flow field as a result of the high degree of spatial variance in soil hydraulic conductivity and other physical characters (Biggar and Nielsen, 1976). A significant feature of preferential flow is the rapid movement of water and solute to significant depths while bypassing a large part of the matrix porespace. As a result, water and solutes may move to far greater depths, and much faster, than would be predicted using averaged flow velocity (Jury and Flühler, 1992). The consequence of the preferential flow is the nonequilibrium transport processes where solute in fast flow regions does not have sufficient time to equilibrate with slowly moving resident water in the bulk of the soil matrix. Therefore, the breakthrough curves of the solute transport through preferential flow often exhibit a sharp front indicating early breakthrough in the rapid flow region, and long tailing as a result of the rate-limited transfer of solutes from the slow flow region to the fast flow region (Jury and Flühler, 1992).

Since macropores in structured soils can conduct water rapidly to deeper soil horizons, while bypassing the denser, less permeable soil matrix (Jarvis, Villholth, and Ulén, 1999), an application of reactive chemicals during dry seasons can cause their rapid transport and to an extensive depth in the soil profile. It is commonly accepted that reactive chemicals move down the soil profile through these preferential flow pathways after application of sewage sludge, wastewater, or smelter residues. Sterckeman et al. (2000) reported that concentrations of Cd, Pb, and Zn increased down to a 2 m depth in soils near smelters. They suggested that earthworm galleries were the main pathways for accelerated particulate metal migration. Conventional laboratory transport studies using column leaching techniques with homogenized soil samples are not capable of capturing the increased mobility of heavy metals through preferential flow. Studies using undisturbed soil samples where the pore structure is kept intact are useful in evaluating the potential of transport through preferential flow. For example, the column leaching experiment of Cd, Zn, Cu, and Pb conducted by Camobreco et al. (1996) showed that the metals were completely retained in homogenized soil columns. In contrast, the preferential flow paths in the undisturbed soil columns allowed metals to pass through the soil profile. Distinctively different characteristics of metal breakthrough curves were observed for the four undisturbed soil columns, reflecting different preferential flow paths. Similarly, Corwin, David, and Goldberg (1999) demonstrated with a lysimeter column study that without accounting for preferential flow, 100% of the applied As was isolated in the top 0.75 m over a 2.5-year period. However, when preferential flow was considered and a representative bypass coefficient was used, about 0.59% of applied As moved beyond 1.5 m. Han et al. (2000) quantified the distribution of several heavy metals in a clay soil grown to pasture which had received poultry waste applications over 25 years. They found significant concentration for Zn up to 1 m when compared to nonamended soil.

8.3 Mobile-Immobile or Two-Region

One of the most popular mechanistic models that is currently used to model the physical nonequilibrium that results from either the presence of aggregates or macropores was presented by van Genuchten and Wierenga (1976), which was based on the scheme developed in petroleum engineering by Coats and Smith (1964). The great simplicity of this model is that one fraction of the soil water Θ^{im} is considered, for transport purposes, to be stagnant. This immobile water may be residing in the intra-aggregate domain, or within the micropores of the matrix. Furthermore, mobile soil water Θ^{m} may be regarded as water residing between the aggregates or flowing in the macropores. This arbitrary division of mobile and immobile soil water fractions is not realistic. Nevertheless, because of its simplicity it provides a convenient analytical tool to assess the influence of mobile-immobile on preferential flow (Clothier, Kirkham, and Mclean, 1992).

For soils dominated by macropores that are connected within a microporous matrix, an earlier-than-expected arrival of solute will occur due to preferential transport along the larger and multiconnected macroporcs (Figure 8.1). This higher velocity transport, coupled with mechanical mixing between the variously sized macropore networks, will result in a breakthrough of inert solute in which the smearing is dominated by hydrodynamic dispersion. The relative role of diffusion from the macropores will depend on the comparative size of the macropores. In the lower limit the pore sizes will be of the same order, and the flow will become uniform since local equilibrium will prevail.

As stated earlier, several experimental studies showed evidence of early breakthrough results and tailing with nonsymmetrical concentration distributions of effluent breakthrough curves (BTCs). Discrepancies from symmetrical or ideal behavior for several solutes led to the concept of solute transfer between mobile and immobile waters. It is commonly postulated that tailing under unsaturated conditions was perhaps due to the fact that larger pores

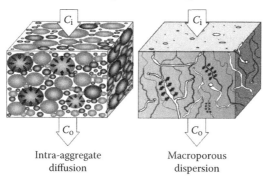

FIGURE 8.1

Local nonequilibrium transport through an aggregated soil where preferential flow is dominated by intraaggregate diffusion (left) and in a macroporous soil where solute breakthrough is controlled by hydrodynamic dispersion. (From B. E. Clothier, M. B. Kirkham, and J . E. Mclean. 1992. *Soil Sci. Soc. Am. J.* 56: 733–736. With permission.)

are eliminated for transport and the proportion of the water that does not readily move within the soil increased. This fraction of water was referred to as stagnant or immobile water. A decrease in water content increases the fraction of air-filled macropores resulting in the creation of additional dead end pores which depend on diffusion processes to attain equilibrium with a displacing solution. However, the conceptual approach of mobile-immobile or two-region behavior is perhaps more intuitively applicable for well-structured or aggregated soils under either saturated or unsaturated flow.

8.3.1 General Formulation

The equations describing the movement for a nonreactive solute through a porous medium having mobile and immobile water fractions are:

$$\Theta^m \frac{\partial C^m}{\partial t} = \Theta^m D \frac{\partial^2 C^m}{\partial x^2} - v^m \Theta^m \frac{\partial C^m}{\partial x} - \alpha (C^m - C^{im}) \tag{8.1}$$

and

$$\Theta^{im} \frac{\partial C^{im}}{\partial t} = \alpha (C^m - C^{im}) \tag{8.2}$$

Equation 8.1 is a modified version of the convection-dispersion equation where D is the hydrodynamic dispersion coefficient in the mobile water region (cm² h⁻¹), Θ^m and Θ^{im} are mobile and immobile water fractions (cm³ cm⁻³), respectively. The terms C^m and C^{im} are the concentrations in the mobile and immobile water (μg cm^{-3}), and v^m is the average pore-water velocity in the mobile region (cm h^{-1}). Also x is depth (cm) and t is time (h). It is also assumed that the immobile water (Θ^{im}) is located inside aggregate pores (interaggregate) where solute transfer occurs by diffusion only. In Equation 8.2 α is a mass transfer coefficient (h^{-1}), which governs the transfer of solutes between the mobile- and immobile-water phases in analogous manner to a diffusion process.

The mobile-immobile concept represented by Equations 8.1 and 8.2 may be generalized for the transport of reactive solutes. Incorporation of reversible and irreversible retention for reactive solutes in Equations 8.1 and 8.2 yields:

$$\Theta^m \frac{\partial C^m}{\partial t} + f\rho \frac{\partial S^m}{\partial t} = \Theta^m D \frac{\partial^2 C^m}{\partial x^2} - v^m \Theta^m \frac{\partial C^m}{\partial x} - \alpha(C^m - C^{im}) - Q^m \quad (8.3)$$

and

$$\Theta^{im} \frac{\partial C^{im}}{\partial t} + (1-f)\rho \frac{\partial S^{im}}{\partial t} = \alpha(C^m - C^{im}) - Q^{im} \quad (8.4)$$

where ρ is soil bulk density (g cm^{-3}). Here the soil matrix is divided into two regions (or sites) where a fraction f is a dynamic or easily accessible region and the remaining fraction is a stagnant or less accessible region (see Figure 8.2). The dynamic region is located close to the mobile phase, whereas the stagnant region is in contact with the immobile phase. Moreover, S^m and S^{im} are the amounts of solutes sorbed in the dynamic and stagnant regions (μg per gram soil), respectively. Also Q^m and Q^{im} are sink (or source) terms associated with the mobile and immobile water regions, respectively. Therefore, Q^m and Q^{im} represent rates of irreversible-type reactions. These terms must be distinguished from S^m and S^{im}, which represent reversible sorbed solutes in the dynamic and stagnant regions, respectively.

The mobile-immobile approach has been successful in describing the fate of several pesticides in soils when linear and Freundlich reversible reactions were considered (van Genuchten, Wierenga, and O'Connor, 1977). However, it is often necessary to include kinetic rather than equilibrium reactions to account for the degradation of pesticides in soils (Rao et al., 1979). The mobile-immobile approach has been successfully used to describe heavy metal transport in soils when adsorption was considered as a Langmuir kinetic along with a first-order irreversible reaction (Selim and Amacher, 1988). The mobile-immobile is approach has had only limited success when extended to describe the transport and exchange of ions in soils for binary (Ca-Mg) and ternary (Ca-Mg-Na) systems (for a review see Selim, 1992).

The mobile-immobile concept is commonly referred to as the two-region model and is regarded as a mechanistic approach where physical

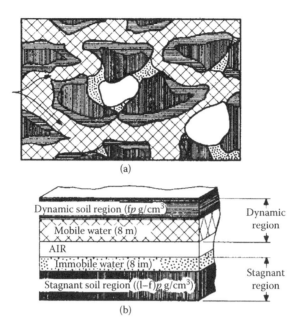

FIGURE 8.2
Schematic diagram of unsaturated aggregated porous medium. (A) Actual model; (B) simplified model. The shading patterns in A and B represent the same regions. (From M. Th. van Genuchten and P. H. Wierenga. 1976. *Soil Sci. Soc. Am. J.* 40: 473–480. With permission.)

nonequilibrium is a controlling mechanism of solute behavior in porous media. On the other hand, chemically controlled heterogeneous reactions are the governing mechanisms for the two-site (equilibrium/kinetic) approaches (see Chapter 5). However, one can show that the two models are analogous mathematically. Therefore, analysis of data sets of effluent results from miscible displacement experiments alone is not sufficient to differentiate between physical and chemical processes as causes for often observed apparent nonequilibrium behavior in soils. The similarity of the two transport models also means that the two formulations can be used in macroscopic and semiempirical manner without having to delineate the exact physical and chemical processes on the microscopic level (Nkedi-Kizza et al., 1984; Selim and Amacher, 1988).

8.3.2 Estimation of α

One major assumption of the mobile-immobile model is that of uniform solute distribution in each water phase. In addition, solute transfer between the two water phases is assumed to follow an empirical first-order diffusion mechanism. An alternative to this approach is to assume spherical aggregate geometry with water within the aggregates as the immobile water phase where solute distribution in the sphere is not considered uniform (Rao et al., 1980a). Moreover, solute diffusion into the aggregates can be governed by Fick's second law, which may be expressed in spherical coordinates as:

$$\frac{\partial C^{im}(rt)}{\partial t} = D_e \left(\frac{\partial^2 C^{im}}{\partial r^2} + \frac{2}{r} \frac{\partial C^{im}}{\partial r} \right) \tag{8.5}$$

where D_e ($= D_o \tau^2$) is an effective molecular diffusion coefficient, D_o is molecular diffusion in water, τ^2 is a tortuosity factor (<1), and r is the radial coordinate in a sphere of diameter a. Here, D_e was assumed to be independent of concentration within the aggregate C^{im}. Average concentration in the sphere can be calculated using:

$$\overline{C}^{im}(t) = \frac{3}{a} \int_0^a r^3 C^{im}(r,t) dr \tag{8.6}$$

As a result, Rao et al. (1980a, 1980b) derived an approximate expression for α assuming spherical aggregates as:

$$\alpha = \left[\frac{D_e \Theta^{im}}{a^2} \right] \alpha^*(t) \tag{8.7}$$

where the parameter α^* is estimated based on the aggregate size a, D_o, t, and F, the fraction of the mobile to total water content ($F = \Theta^m / \Theta$). As a result, this α or $\alpha(t)$ in Equation 8.7 is time dependent and approximates the diffusion process in a sphere. In an earlier attempt to arrive at an expression for an overall dispersion D for nonreactive solute transport in spherical aggregates, Passioura (1971) approximated the overall D for soils composed of spherical aggregates as:

$$D = D^m F + \frac{(1-F)a^2 v^2}{15 D_e} \tag{8.8}$$

with the constraint

$$\frac{(1-F)D_e L}{a^2 v^2} > 0.3 \tag{8.9}$$

where L is solute transport length. As evidenced by Equation 8.8, when physical nonequilibrium is dominant, the overall D increases with increasing velocity and aggregate sizes. Equation 8.8 was extended to include a reactive solute with a retardation factor R (van Genuchten and Dalton, 1986; Parker and Valocchi, 1986):

$$D = D^m F + \frac{(1-F)a^2v^2(R^{im})^2}{15D_eR^2} \tag{8.10}$$

where R^{im} is the retardation factor associated with the immobile phase, R is an overall retardation factor ($\Theta R = \Theta^{im} R^{im} + \Theta^m R^m$). Similar effective dispersion coefficients were obtained for rectangular aggregates with half width of a_l as:

$$D = D^m F + \frac{(1-F)a_l^2 \, v^2(R^{im})^2}{3D_eR^2} \tag{8.11}$$

An overall D can also be derived from empirical first-order mass transfer where uniform solute distribution in the immobile water phase may be assumed. This was carried out by De Smedt and Wierenga (1984), who derived an expression for D for nonreactive solutes for long columns as:

$$D = D^m F + \frac{\Theta(1-F)^2 v^2}{\alpha} \tag{8.12}$$

and for reactive solutes (van Genuchten and Dalton, 1986) as:

$$D = D^m F + \frac{(1-F)a_l^2v^2(R^{im})^2}{\alpha R^2} \tag{8.13}$$

Comparing Equations 8.8 and 8.12 an equivalent first-order transfer coefficient (α_t) for spherical aggregates is thus obtained:

$$\alpha_t = \frac{15D_e(1-F)\,\Theta}{a^2} \tag{8.14}$$

This equation can also be used using moment analysis (Valocchi, 1985) and has been used in solute transport (Selim, Schulin, and Flühler, 1987; Selim and Amacher, 1988). Similar α expressions were obtained for a rectangular aggregate as:

$$\alpha = \frac{3D_e(1-F)\,\Theta}{a_l^2} \tag{8.15}$$

As a result, a more general formulation for estimating α_c is

$$\alpha_c = \frac{nD_e\,(1-F)\,\Theta}{a_e^2} \tag{8.16}$$

where n is a geometry factor and a is an average effective diffusion length. Details can be found in van Genuchten and Dalton (1986) and van Genuchten (1985). Equation 8.14 has been used to estimate solute transport in porous media (Selim and Gaston, 1990; Goltz and Roberts, 1988).

8.3.3 Estimation of Θ^m and Θ^{im}

A common way of estimating the mobile and immobile water content is by use of curve fitting of tracer breakthrough results (Li and Ghodrati, 1994; van Genuchten and Wierenga, 1977). De Smedt and Wierenga (1984) found $\Theta^{im} = 0.8530$ is applicable for unsaturated glass beads with diameters in the neighborhood of 100 μm. Alternatively, a direct method of estimating Θ^m and Θ^{im} is by measuring the soil water content at some arbitrary water tension (ψ). Smettem and Kirkby (1990) used water content at $\psi = 14$ cm as the matching point between the interaggregate (macro-) and the intraaggregate (micro-) porosity by examining the $\psi - \Theta$ soil moisture characteristic curve. Jarvis, Bergstrom, and Dik (1991) estimated macroporosity from specific yield under water tension of 100 cm. Other water tensions used to differentiate macropores from micropores are 3 cm (Luxmoore, 1981), 10 cm (Wilson, Jardine, and Gwo, 1992), 20 cm (Selim, Schulin, and Flühler, 1987), and 80 cm (Nkedi-Kizza et al., 1982). A list of water tensions used by different authors was provided by Chen and Wagenet (1992). The equivalent diameters at these water tensions range from 10 to 10,000 μm. Another experimental measurement of Θ^{im} is based on the following mass balance equation:

$$\Theta R = \Theta^m\,R^m + \Theta^{im}\,C^{im} \tag{8.17}$$

When α is small enough to assume $C^{im} = 0$ and C_o (input concentration) at certain infiltration time t, the approximate equation is obtained:

$$\Theta^m = \Theta\frac{C}{C^m} = \Theta\frac{C}{C_o} \tag{8.18}$$

Applications of this method can be found in Clothier, Kirkham, and Mclean (1992) and Jaynes et al. (1995). By assuming that a tracer concentration in the mobile water phase (C^m) equals input solution concentration (C_o), Jaynes et al. (1995) derived the following formula from Equations 8.2 and 8.18:

$$\ln\left(1-\frac{C}{C_o}\right) = -\frac{\alpha t}{\Theta^{im}} + \ln\left(\frac{\Theta^{im}}{\Theta}\right) \tag{8.19}$$

α and Θ^{im} can be estimated by plotting $\ln(1 - C/C_o)$ versus application (infiltration) time. However, the assumption of $C^{im} = C_o$ associated with this method is questionable and may not be correct as long as α is not equal to 0. A slightly different approach was used by Goltz and Roberts (1988) to estimate the fraction of mobile water as the ratio of velocity calculated from hydraulic conductivity to the velocity measured from tracer experiment.

8.3.4 Experimental Evidence

Traditionally, solute transport through structured subsurface media is conceptualized as both a rapid transport process occurring within preferential flow paths (fractures and macropores) and a diffusive mass transfer process that occurs between the rapid flow region and a stagnant region (micropores), for example, the mobile-immobile concept (Nkedi-Kizza et al., 1982; Parker and van Genuchten, 1984). The disparity in solute mobility in the different pore regions results in concentration gradients between regions, that is, physical nonequilibria, whereby diffusive mass transfer serves to reestablish equilibrium.

Physical nonequilibrium is often investigated by employing a miscible displacement technique with the continuous injection of a nonreactive tracer and the observation of an asymmetric breakthrough curve (Smettem, 1984; Schulin et al., 1987; Buchter et al., 1995). Under certain conditions, nonequilibrium cannot be readily distinguished when analyzing data obtained with the typical continuous flow column experiment (Brusseau et al., 1989). There are several tracer studies in the literature that illustrate deviations of experimental breakthrough results from predictions based on the convection-dispersion equation for homogeneous porous media where physical nonequilibrium was not considered. The tritium BTCs shown in Figures 8.4 and 8.5 illustrate lack of equilibrium conditions and is commonly considered evidence of physical nonequilibria.

Another technique known as *flow interruption* or simply stop flow has been frequently used to assess nonequilibrium processes in miscible displacement experiments. This method is based on the "interruption test," which was developed in chemical engineering to distinguish between systems that were controlled by intraaggregate diffusion and those controlled by film diffusion. Murali and Aylmore (1980) are perhaps the earliest to implement the flow interruption technique during the displacement of several ions through a finely ground soil.

The flow interruption technique has been used to a much lesser extent to assess the significance of physical nonequilibria during solute transport (Koch and Flühler, 1993). Hu and Brusseau (1995), and Reedy et al. (1996) among others, implemented the flow interruption method to quantify physical nonequilibria processes during displacement of a conservative mobile dye tracer or bromide in packed columns. Their results suggested that distinct physical nonequilibria mechanisms influenced the mobility and shape of the breakthrough results. Such examples where nonreactive or nonsorbing tracers were used showed the significance of diffusion during solute transport in aggregated media.

The mobile-immobile approach described above, which is commonly referred to as the two-region model, is regarded as a mechanistic approach where physical nonequilibrium causes solute transfer between regions. Flow interruption can be accounted for in the above formulation by simply assuming the flux v as zero and $D = D_o$ during interruption, where D_o is the effective molecular diffusion coefficient. As a result, Equation 8.1 is reduced to the commonly known diffusion equation:

$$\Theta^m \frac{\partial C^m}{\partial t} = \Theta^m D_o \frac{\partial^2 C^m}{\partial x^2} - \alpha\,(C^m - C^{im})$$

(8.20)

Therefore, during flow interruption, we assume that transport of nonreactive solutes can be described by molecular diffusion within the mobile water phase. Thus, during flow interruption, the exchange of solute between mobile and immobile phases is assumed to follow the simple mass transfer in Equation (8.20).

The examples presented here to illustrate physical nonequilibria are those of the work of Reedy et al. (1996) where transport of the nonreactive tracer bromide (Br) in packed columns was investigated. Reedy et al. (1996) carried out a series of experiments to examine the time dependency of the diffusive mass transfer process by imposing flow interruptions of increasing duration (0.25 to 7 d) on tracer experiments conducted at similar fluxes (Figure 8.6). Stop flow or simply flow interruption during tracer pulse application resulted in decreased Br concentrations, whereas flow interruption during tracer leaching or displacement resulted in increased Br concentrations when flow was reinitiated (Figure 8.6). These observed concentration perturbations resulting from flow interruption are indicative of solute diffusion between the two regions (fracture and matrix) of the porous medium. Conditions of preferential flow create concentration gradients between pore domains, resulting in diffusive mass transfer between the regions. Concentration perturbations observed in Figure 8.6 were driven by solute concentration gradients established between pore regions, that is, physical nonequilibria. Upon stop flow, the decrease in relative concentration indicates that solute diffusion is occurring from the larger, more conductive pores, into the smaller pores. During tracer displacement or leaching, Br concentrations within the preferred flow paths are lower than those within the matrix. The relative concentration increase after the second interruption indicates solute diffusion from smaller pore regions back into larger pore regions (Figure 8.6).

The extent of concentration perturbations can be observed in the Br BTC with increased interruption time. As the duration of stop flow increased, the concentration perturbation increased, which suggests that the diffusive process was far removed from equilibrium after short flow interruption durations, and that a significant time period was required to alleviate the concentration gradient that was established between large and small pore domains during flowing conditions. In addition, one value of the parameter α was used to obtain these simulations for the different columns having increasing stop flow duration. This finding is significant and lends credence to physical nonequilibria in soils.

The concentration perturbations that result from flow interruptions place a constraint on the model simulations such that a limited range in certain parameter values was acceptable for matching the observed data. As shown in Figure 8.7, model simulations are highly dependent on the parameter F where the solid curve represents best simulation of the experimental data. Large values of F described the rapid breakthrough of the bromide tracer as well as the ascending and descending limbs of the BTC. However, the concentration perturbation at the stop flow was completely absent. Smaller values of F well described the influence of flow interruption. However, the model grossly overestimated the rapid breakthrough of the tracer. Likewise, only a limited range of α values could satisfactorily describe the experimental BTC.

Model sensitivity to a range of the mass transfer coefficient α in describing Br breakthrough results is illustrated in Figure 8.8. The solid curve represents best simulation of the experimental data. Small values of α overpredicted the rapid breakthrough of tracer and the concentration perturbation that resulted from flow interruption. Large values of α described a majority of the BTCs; however, the simulated concentration perturbation was absent. If stop flow interruption was not accounted for in the model to describe the experimental data, the resulting simulations would be less than optimum. Flow interruption places a constraint on the modeling effort that is advantageous for predictions of observed transport data.

8.4 A Second-Order Approach

In the previous chapter, the second-order reactions associated with sites 1 and 2 were considered as kinetically controlled, heterogeneous chemical retention reactions (Rubin, 1983). One can assume that these processes are predominantly controlled by surface reactions of adsorption and exchange. In this sense, the second-order model is along the same lines as the earlier two-site model of Selim et al. (1976) and Cameron and Klute (1977). Another type of two-site model is that of Villermaux (1974), which is capable of describing BTCs from chromatography columns having two concentration maxima.

In this section, the second-order concept is invoked where processes of solute retention were controlled by two types of reactions; namely, chemically controlled heterogeneous reaction and the other physically controlled reaction (Rubin, 1983). The chemically controlled heterogeneous reaction was considered to be governed according to the second-order approach. In the meantime, the physically controlled reaction is chosen to be described by diffusion or mass transfer based on the mobile-immobile concept (Coats and Smith, 1964; van Genuchten and Wierenga, 1976). A comparison between the mobile-immobile concept and that of the two-site approach indicates that the dynamic and stagnant regions for solute retention are analogous to sites 1 and 2 of the two-site concept. Nkedi-Kizza et al. (1984) presented a detailed discussion on the equivalence of the mobile-immobile and the equilibrium-kinetic two-site models.

An important feature of the second-order approach is that an adsorption maximum (or capacity) is assumed. This maximum represents the total number of adsorption sites per unit mass or volume of the soil matrix. It is also considered an intrinsic property of an individual soil and is thus assumed constant. In previous two-region models, a finite number of sites has not been specified and thus an adsorption maximum is never attained (e.g., van Genuchten and Wierenga, 1976; Rao et al., 1979; Brusseau et al., 1989; van Genuchten and Wagenet, 1989). Specifically, in Equations 8.3 and 8.4, the dimensionless term f denotes the ratio or fraction of dynamic or active sites to the maximum or total adsorption sites S_{max} (μg/g soil). In addition, the terms (MS^m/Mt) and (MS^{im}/Mt) represent the rates of reversible heavy metal reactions between C in soil solution and that present on matrix surfaces in the mobile (or dynamic) and the immobile regions, respectively. Moreover, irreversible reactions of heavy metals were incorporated in this model as may be seen by the inclusion of the sink terms Q^m and Q^{im} (μg cm^{-3} h^{-1}) in Equations 8.3 and 8.4, respectively. It is assumed that irreversible retention or removal from solution will occur separately in the mobile and immobile water regions. However, the governing mechanism of retention for each region was assumed to follow first-order-type reactions. Specifically, it is assumed that irreversible retention for the mobile and immobile regions are considered in this transport model as follows:

$$Q^m = \Theta^m \, k_s \, C^m \tag{8.21}$$

$$Q^{im} = \Theta^{im} \, k_s \, C^{im} \tag{8.22}$$

where k_s is an irreversible rate coefficient (h^{-1}) common to both regions. These formulations for the sink terms do not occur elsewhere in the literature and were first proposed by Selim and Amacher (1988). Since soil matrix surfaces may behave in a separate manner to heavy metal retention, it is conceivable that the rate of irreversible reactions in the mobile region is characteristically different from that for the mobile region. One way to achieve this is to distinguish between the rate coefficient (k_s) controlling the reaction for the two regions, for example, k^m_s and k^{im}_s in Equations 8.5 and 8.6, respectively. However, such a distinction in reaction coefficients was not incorporated in this model and a common parameter k_s was thus used.

The retention mechanism associated with the mobile and immobile phases of Equations 8.3 and 8.4 was considered as an equilibrium linear sorption by van Genuchten and Wierenga (1976) and was extended to the nonlinear or Freundlich type by Rao et al. (1979). Recently, multiple ion retention expressed on the basis of ion-exchange equilibrium reactions was successfully incorporated into the mobile-immobile model by van Eijkeren and Lock (1984) and Selim, Schulin, and Flühler (1987). Here, it is considered that reversible solute reactions are governed by the second-order kinetic approach. Specifically, the rates of reaction for S^m and S^{im} were considered as (Selim and Amacher, 1988):

$$\rho \frac{\partial S^m}{\partial t} = \Theta^m k_1 \varphi^m C^m - \rho k_2 S^m \tag{8.23}$$

and

$$\rho \frac{\partial S^{im}}{\partial t} = \Theta^{im} k_1 \varphi^{im} C^{im} - \rho k_2 S^{im} \tag{8.24}$$

where k_1 and k_2 are forward and backward rate coefficients (day^{-1}), respectively. Here φ^m and φ^{im} represent the vacant or unfilled sites (μg per gram soil) within the dynamic and the stagnant regions, respectively. In addition, the terms φ^m and φ^{im} can be expressed as:

$$\varphi^m = S^m_{max} - S^m = f \ S_{max} - S^m \tag{8.25}$$

$$\varphi^{im} = S^{im}_{max} - S^{im} = (1 - f) \ S_{max} - S^{im} \tag{8.26}$$

where S_{max}, S_{max}^m, and S_{max}^{im} are the total number of sites in the soil matrix, total sites in the dynamic region, and the total in the less accessible region (mg/kg soil), respectively. These terms are related by:

$$S_{max} = S^m_{max} + S^{im}_{max} \tag{8.27}$$

We also assume S_{max} to represent the combined total of occupied and unoccupied sites, that is, maximum adsorption capacity of an individual soil, and it is regarded as an intrinsic property of the soil.

An important feature of the second-order retention approach (Equations 8.23 and 8.24) is that similar reaction rate coefficients (k_1 and k_2) associated with the dynamic and stagnant regions were chosen. Specifically, it is assumed that the retention mechanism is equally valid for the two regions of the porous medium. A similar assumption was made by van Genuchten and Wierenga (1976) for equilibrium linear and Freundlich-type reactions and by Selim, Schulin, and Flühler (1987) for selectivity coefficients for homovalent ion-exchange reactions. Specifically, as $t \to \infty$, that is, when both the dynamic (or active) sites and the sites in the stagnant region achieve local equilibrium, Equations 8.23 and 8.24 yield the following expressions. For the active sites associated with the mobile region:

$$\Theta^m k_1 \varphi^m C^m - \rho k_2 S^m = 0 \qquad as \quad t \to \infty \qquad (8.28)$$

or

$$\frac{S^m}{\varphi^m C^m} = \frac{\Theta^m k_1}{\rho k_2} = \omega_1 \qquad as \quad t \to \infty \qquad (8.29)$$

and for the sites associated with the immobile region we have:

$$\Theta^{im} k_1 \varphi^{im} C^{im} - \rho k_2 S^{im} = 0 \qquad as \quad t \to \infty \qquad (8.30)$$

or

$$\frac{S^{im}}{\varphi^{im} C^{im}} = \frac{\Theta^{im} k_1}{\rho k_2} = \omega_2 \qquad as \qquad t \to \infty \qquad (8.31)$$

Here ω_1 and ω_2 represent equilibrium constants for the retention reactions associated with the mobile and immobile regions, respectively. The formulation of Equations 8.30 and 8.32 are analogous to expressions derived for the second-order two-site model discussed previously. In this sense, the equilibrium constants ω_1 and ω_2 resemble the Langmuir coefficients, with S_{max} as the maximum sorption capacity (Selim and Amacher, 1988). These equilibrium constants are also analogous to the selectivity coefficients associated with ion exchange reactions (see Selim, Schulin, and Flühler, 1987).

The dimensionless forms of Equations 8.3, 8.4, 8.23, and 8.24 are

$$\Omega f \frac{\partial s^m}{\partial T} + \mu \frac{\partial c^m}{\partial T} = \frac{\mu}{P} \frac{\partial^2 c^m}{\partial X^2} - \frac{\partial c^m}{\partial X} - \bar{\alpha}(c^m - c^{im}) - \mu k_s c^m \qquad (8.32)$$

$$\frac{\partial c^{im}}{\partial T} + \Omega \frac{1-f}{1-\mu} \frac{\partial s^{im}}{\partial T} = \bar{\alpha}(c^m - c^{im}) - k_s c^{im} \qquad (8.33)$$

$$\frac{\partial s^{im}}{\partial T} = \mu k_1 \Phi^m c^m - k_2 s^m \qquad (8.34)$$

$$\frac{\partial s^{im}}{\partial T} = (1-\mu)k_1 \Phi^{im} c^{im} - k_2 s^{im} \qquad (8.35)$$

where

$$c^m = \frac{C^m}{C_0}, \qquad c^{im} = \frac{C^{im}}{C_0} \tag{8.36}$$

$$s^m = S^m/S_{max}, \; s^{im} = S^{im}/S_{max}, \; \Phi^m = \varphi^m/S_{max}, \; \Phi^{im} = \varphi^{im}/S_{max}, \tag{8.37}$$

$$P = vL/D\Theta, \tag{8.38}$$

$$= \alpha L/v(1 - \mu), \tag{8.39}$$

$$\mu = \Theta^m/\Theta, \tag{8.40}$$

$$X = x/L, \tag{8.41}$$

$$T = vt/L\Theta, \tag{8.42}$$

where T is dimensionless time equivalent to the number of pore volumes leached through a soil column of length L, and P is the Peclet number (Brenner, 1962). In addition, we have defined:

$$\Omega = S_{max}\rho/C_o\Theta, \tag{8.43}$$

$$\kappa_s = k_s\Theta L/v, \tag{8.44}$$

$$\kappa_1 = k_1 \, \Theta^2 C_o L/\rho v, \tag{8.45}$$

$$\kappa_2 = k_2\Theta L/v. \tag{8.46}$$

Here, κ_s, κ_1, and κ_2 are dimensionless kinetic rate coefficients that incorporate v and L.

8.4.1 Initial and Boundary Conditions

The corresponding initial and boundary conditions associated with the second-order mobile-immobile model can be expressed as:

$$C^m = C^{im} = C_i \quad (t = 0, 0 < x < L) \tag{8.47}$$

$$S^m = S^{im} = S_i \quad (t = 0, 0 < x < L) \tag{8.48}$$

$$vC_o = vC^m - \Theta^m D\frac{\partial C^m}{\partial x} \quad (x = 0, t < t_p) \tag{8.49}$$

$$0 = vC^m - \Theta^m D \frac{\partial C^m}{\partial x} \qquad (x = 0, t > t_p) \qquad (8.50)$$

$$\frac{\partial C^m}{\partial x} = 0 \qquad (x = L, t \geq 0) \qquad (8.51)$$

These conditions are similar to those described earlier for the transport of a solute pulse (input) in a uniform soil having a finite length L where a steady water flux v was maintained constant. The soil column is considered as having uniform retention properties as well as having uniform ρ and Θ. It is further assumed that equilibrium conditions exist between the solute present in the soil solution of the mobile water phase (i.e., interaggregate) and that present in the immobile (or interaaggregate) phase. This necessary condition is expressed by Equations 8.47 and 8.48. Uniform initial conditions were assumed along the soil column. It is assumed that an input heavy metal solution pulse having a concentration C_o was applied at the soil surface for a time duration t_p and was then followed by a solute-free solution. As a result, at the soil surface, the third-type boundary conditions were those of Equations 8.49 and 8.50. In a dimensionless form, the boundary conditions can be expressed as:

$$1 = c^m - \Theta^m D \frac{\partial c^m}{\partial x} \qquad (X = 0, T < T_p) \qquad (8.52)$$

$$0 = c^m - \Theta^m D \frac{\partial c^m}{\partial x} \qquad (X = 0, T > T_p) \qquad (8.53)$$

and at $x = L$, we have

$$\frac{\partial c^m}{\partial x} = 0 \qquad (X = L, T > 0) \qquad (8.54)$$

where T_p is dimensionless time of input pulse duration of the applied solute and represents the amount of applied pore volumes of the input solution.

8.4.2 Sensitivity Analysis

Figures 8.9 through 8.12 are examples of simulated BTCs to illustrate the sensitivity of the proposed second-order reaction, when incorporated into the mobile-immobile concept, to various model parameters. As shown, several features of the mobile-immobile concept dominate the behavior of solute transport and thus the shape of simulated BTCs. For this reason, we restrict the discussion here to the influence of parameters pertaining to the proposed second-order mechanism. Specifically, the influence of κ_1 and κ_2, and ω on solute retention were examined. Other parameters such as D and v have been rigorously examined in earlier studies (Coats and Smith, 1964; van Genuchten and Wierenga, 1976).

For the simulations shown in Figures 8.2 to 8.6, initial conditions, volume of input pulse, and model parameters were identical to those used previously for the second-order two-site model, where $C_i = S_i = 0$ within the mobile and immobile regions. Specifically, the parameters chosen were: $L = 10$ cm, $D = 1$ cm^2 h^{-1}, $\rho = 1.2$ g cm^{-3}, $f = 0.50$, $\Theta = 0.40$ cm^3 cm^{-3}, $\mu = \Theta^m/\Theta = 0.5$, $C_o = 100$ mg L^{-1}, and a Peclet number $P = 25$. Moreover, unless otherwise stated, the values selected for the dimensionless parameters κ_1, κ_2, κ_s, Ω, were 1, 1, 0, 5, and 1, respectively. It is assumed a solute pulse was applied to a fully water-saturated soil column initially devoid of a particular heavy metal of interest. In addition, a steady water flow velocity (v) was maintained constant with a Peclet number P of 25. The length of the pulse was assumed to be three pore volumes, which was then followed by several pore volumes of metal-free solution.

The influence of the reaction rate coefficients on the shape of the BTCs is illustrated in Figure 8.9. Here the values of κ_1 and κ_2 were varied simultaneously provided that κ_1/κ_2 (and ω_1 and ω_2) remained invariant. For the nonreactive case ($\kappa_1 = \kappa_2 = 0$), the highest effluent peak concentration and least tailing was observed. As the rate of reactions increased simultaneously, solute peak concentrations decreased and excessive tailing of the BTCs was observed. However, the arrival time or the location of peak concentration was not influenced by increasing the rates of reactions.

The effect of increasing values of the equilibrium constant ω, which represents the ratio κ_1/κ_2, on the shape of BTCs is shown in Figure 8.3. Here a constant value for κ_2 of 1 was chosen, whereas κ_1 was allowed to vary. For all BTCs shown in Figure 8.10, the values of ω_1 and ω_2 were equal (since $\mu = 0.5$). As a result we refer to simply ω rather than ω_1 and ω_2. The results indicate that as the forward rate of reaction (κ_1) increased, an increase in solute retardation or a right shift of the BTCs was observed. This shift of the BTCs was accompanied by an increase in solute retention (i.e., a decrease of the amount of solute in the effluent, based on the area under the curve) and a lowering of peak concentrations. Similar behavior was observed for the influence of the dimensionless transfer coefficient ω on the shape of the BTCs as may be seen from the BTCs in Figure 8.4. For exceedingly large values (>2), the diffusion between the mobile and immobile phases became more rapid. Therefore, equilibrium condition between the two phases is nearly attained (Valocchi, 1985).

Figure 8.12 shows BTCs of a reactive solute for several values of Ω. The figure indicates that the shape of the BTCs is influenced drastically by the value of Ω. This is largely due to the nonlinearity of the proposed second-order retention mechanism. As given by Equation 8.25, Ω represents the ratio of total sites (S_{max}) to input (pulse) solute concentration (C_o). Therefore, for

Mobile-Immobile Multi-Reaction Model (MIM-MRM)

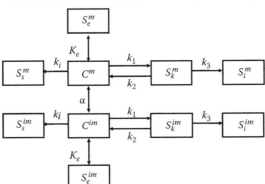

FIGURE 8.3
A combined physical and chemical nonequilibrium model based on the mobile-immobile model and multireaction retention approaches, respectively.

small values of Ω (e.g. $\Omega = 0.1$), the simulated BTC is very similar to that for a nonretarded solute due to the limited number of sites (S_{max}) in comparison to C_o. In contrast, large values of Ω resulted in BTCs that indicate increased retention as manifested by the right shift of peak concentration of the BTCs. In addition, for high values of Ω, extensive tailing as well as an overall decrease of effluent concentration was observed. The influence of the parameter f, which represents the fraction of active or dynamic sites within the mobile region to the total number of sites, on the behavior of solute retention and transport is shown in Figure 8.13 for several values of f. There are

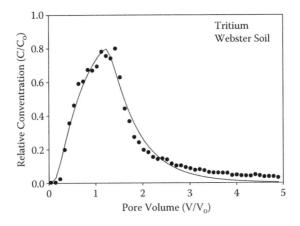

FIGURE 8.4
Breakthrough of tritium in a Webster soil column. Simulations are based on the convection-dispersion equation where equilibrium is assumed.

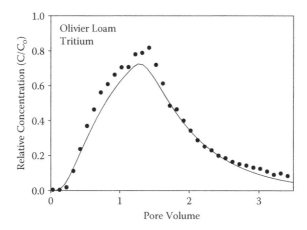

FIGURE 8.5

Breakthrough of tritium in an Olivier soil. Simulations are based on the convection-dispersion equation where equilibrium is assumed.

similar features between these BTCs and those illustrated in the previous figures. For $f = 1$, all the sites are active sites and thus there is no solute retention by the sites present within the immobile region (i.e., stagnant sites). As the contribution of the stagnant sites increases (or f decreases), the shape of the BTCs becomes increasingly less kinetic, with significant increase of the tailing of the desorption side of the BTCs.

In the BTCs shown in Figures 8.2 through 8.6, the irreversible retention mechanism for heavy metal removal (via the sink term) was ignored. The influence of the irreversible kinetic reaction (e.g., precipitation) is a straightforward one and is thus not shown. This is manifested by the lowering of solute concentration for the overall BTC for increasing values of k_s. Since a first-order reaction was assumed, the lowering of the BTC is proportional to the solution concentration. The influence of other parameters on the behavior of solute in soils with the second-order mobile-immobile model, such as P, D, and v, have been studied elsewhere (van Genuchten and Wierenga, 1976).

8.4.3 Examples

The capability of the second-order mobile-immobile model to describe the transport of heavy metals in soils was examined for hexavalent chromium (Cr(VI)) by Selim and Amacher (1988). Their Cr(VI) miscible displacement results and model predictions for three soils are shown in Figures 8.14 to 8.16. To obtain the predictions, several assumptions were necessary for the estimation of model parameters. The sorption maximum (S_{max}) was estimated from kinetic adsorption isotherms. In addition, the ratio of the mobile to

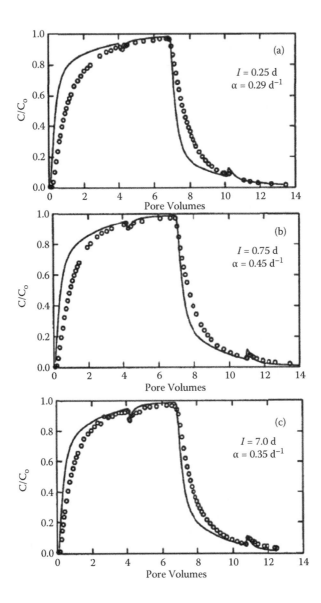

FIGURE 8.6
Breakthrough of bromide in a structured soil with flow interruptions of 0.25, 0.75, and 7 d durations. Simulations are based on the mobile-immobile model.

total water contents (Θ^m/Θ) was estimated based on soil-moisture retention relations for each soil. Selim and Amacher (1988) also assumed that the fraction of sites f is the same as the relative amount of water in the two regions, that is, $f = \mu$ (= Θ^m/Θ). Such an assumption was made because independent measurement of f is not available (van Genuchten and Dalton, 1986). Selim,

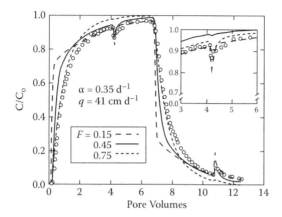

FIGURE 8.7
Breakthrough of bromide in a structured soil. Simulations are based on the mobile-immobile model for different values of the fraction of the mobile water (F).

Schulin, and Flühler (1987) successfully used such estimates of f for a well-aggregated soil. Estimates for α were obtained using:

$$\alpha = 15 D^a \Theta \frac{1-\mu}{a^2} \tag{8.55}$$

where $\mu = \Theta^m / \Theta$ and D^a is the molecular diffusion coefficient in a soil consisting of uniform aggregates of radius a. The above equation was derived by Parker and Valocchi (1986) and is based on time-moment analysis for

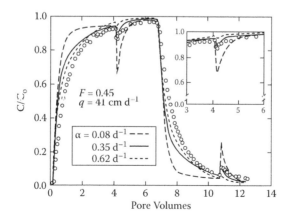

FIGURE 8.8
Breakthrough of bromide in a structured soil. Simulations are based on the mobile-immobile model for different values of the transfer coefficient (α).

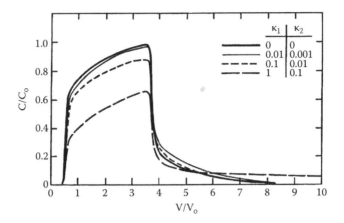

FIGURE 8.9
Effluent concentration distributions for different values of κ_1 and κ_2 using the SOMIM model. (From H. M. Selim and M. C. Amacher. 1997. *Reactivity and Transport of Heavy Metals in Soils.* Boca Raton, FL: CRC Press. With permission.)

spherical diffusion and for first-order kinetic (mobile-immobile) models. Estimates for α are based on the assumption of average aggregate sizes of 0.01, 0.01, and 0.005 cm for Windsor, Olivier, and Cecil soils used, respectively. In addition, D^a for Cr(VI) diffusion was assumed to be 10^{-9} cm^2 sec^{-2} for all three soils. Barber (1984) compiled diffusion coefficients for a number of ions in soils with values for phosphate ($H_2PO_4^-$) ranging from 10^{-8} to 10^{-11} cm^2 sec^{-1} (for water $D^a = 10^{-6}$ cm^2 sec^{-1}). Since chromate and phosphate

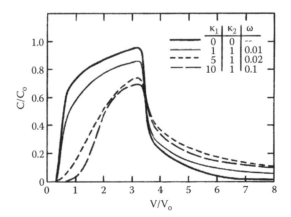

FIGURE 8.10
Effluent concentration distributions for different values of the parameter ω using the SOMIM model. Values of κ_1 and κ_2 are also shown. (From H. M. Selim and M. C. Amacher. 1997. *Reactivity and Transport of Heavy Metals in Soils.* Boca Raton, FL: CRC Press. With permission.)

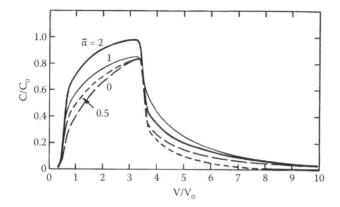

FIGURE 8.11
Effluent concentration distributions for different values of the dimensionless mass transfer parameter (α) of the SOMIM model. (From H. M. Selim and M. C. Amacher. 1997. *Reactivity and Transport of Heavy Metals in Soils*. Boca Raton, FL: CRC Press. With permission.)

have somewhat similar behavior in soils, one can assume that diffusion coefficients for chromate would be equivalent to those for phosphate. These values of D^a and a were also used to estimate the hydrodynamic dispersion coefficient (D) in the mobile-water region using (Parker and Valocchi, 1986):

$$D = \frac{1}{\mu}\left(D^e - \frac{1-\mu}{15}a^2 v^2 D^a\right)$$

(8.56)

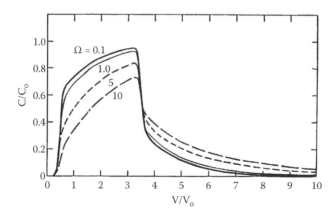

FIGURE 8.12
Effluent concentration distributions for different values of the parameter Ω of the SOMIM model. (From H. M. Selim and M. C. Amacher. 1997. *Reactivity and Transport of Heavy Metals in Soils*. Boca Raton, FL: CRC Press. With permission.)

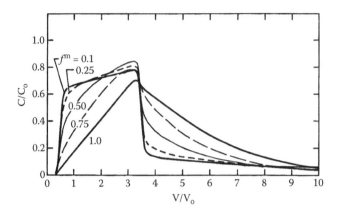

FIGURE 8.13
Effluent concentration distributions for different values of the fraction of active sites f^m of the SOMIM model. (From H. M. Selim and M. C. Amacher. 1997. *Reactivity and Transport of Heavy Metals in Soils*. Boca Raton, FL: CRC Press. With permission.)

where v is the pore water velocity (v/Θ^m). Values for the dispersion coefficient D^e used in the above equation were averages of those obtained from BTCs of tritium and chloride-36 tracers. Selim and Amacher (1988) found that attempts to utilize tracer data sets from tritium and chloride-36 for parameter estimation of α and D were not successful. The values obtained using least squares parameter optimization were inconsistent and ill-defined due to large parameter

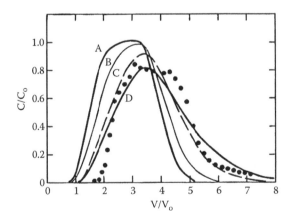

FIGURE 8.14
Effluent concentration distributions for Cr in Olivier soil. Curves A, B, C, and D are predictions using the second-order mobile-immobile model with batch rate coefficients for C/C_o of 25, 10, 5, and 1 mg/L, respectively. (From H. M. Selim, and M. C. Amacher. 1997. *Reactivity and Transport of Heavy Metals in Soils*. Boca Raton, FL: CRC Press. With permission.)

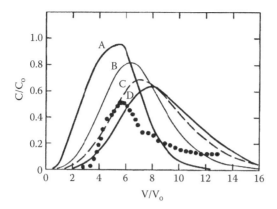

FIGURE 8.15
Effluent concentration distributions for Cr in Windsor soil. Curves A, B, C, and D are predictions using the second-order mobile-immobile model with batch rate coefficients for C/C_o of 25, 5, 2, and 1 mg/L, respectively. (From H. M. Selim and M. C. Amacher. 1997. *Reactivity and Transport of Heavy Metals in Soils.* Boca Raton, FL: CRC Press. With permission.)

standard errors and were often physically unrealistic (e.g., a retardation factor $R \gg 1$). Perhaps these results are due to local equilibrium conditions between the mobile and immobile regions for the two tracers (Rubin, 1983; Parker and Valocchi, 1986).

Values for the rate coefficients used in the second-order mobile-immobile model were those calculated from batch kinetic results. Selim and Amacher

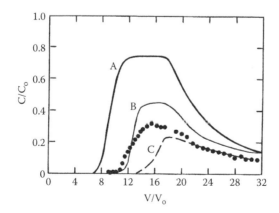

FIGURE 8.16
Effluent concentration distributions for Cr in Cecil soil. Curves A, B, and C are predictions using the second-order mobile-immobile model with batch rate coefficients for C/C_o of 25, 10, and 5 mg/L, respectively. (From H. M. Selim and M. C. Amacher. 1997. *Reactivity and Transport of Heavy Metals in Soils.* Boca Raton, FL: CRC Press. With permission.)

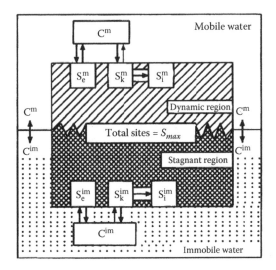

FIGURE 8.17
Schematic diagram of a modified mobile-immobile (two-region) concept. (From H. M. Selim and M. C. Amacher. 1997. *Reactivity and Transport of Heavy Metals in Soils.* Boca Raton, FL: CRC Press. With permission.)

(1988) used k_1, k_2, and k_s values from a three-parameter version of the second-order model described earlier. Predicted BTCs were obtained using different sets of batch rate coefficients due to their strong dependence on input concentrations (C_o). Closest predictions to experimental Cr(VI) measurements were obtained from batch rate coefficients at low C_o values ($C_o < 10$ mg L^{-1}). Moreover, the use of rate coefficients at higher values of C_o resulted in decreased tailing and reduced retardation of the BTCs. These observations are consistent with the second-order two-site approach discussed in the previous chapter. Overall, predictions of measured Cr(VI) using this model may be considered adequate. However, it is conceivable that a set of applicable rate coefficients over the concentration range for Cr(VI) transport experiments cannot be obtained simply by use of the batch procedure described in this study. In addition, several parameters used in model calculations were estimated and not measured, for example, Θ^m, α, and D. The fraction of active sites f was not estimated; rather it was assumed equal to the mobile water fraction (Θ^m/Θ). Amacher and Selim (1988) argued that it is likely that improved model predictions could be obtained if such parameters could be measured independently. They also postulated that other possible factors responsible for these predictions may be due in part to lack of nonequilibrium conditions between the mobile and immobile fractions (Valocchi, 1985; Parker and Valocchi, 1986).

8.5 A Modified Two-Region Approach

Due to the uncertainty of obtaining independent measurement for the fraction of sites f, Selim and Ma (1995) reexamined the original assumption of the mobile-immobile approach within the scope of the second-order formulation described above. Figure 8.17 presents a multiple processes based on a combined physical and chemical nonequilibrium on the concepts of mobile-immobile model and multireaction models. Because there is no practical approach for separating the chemical reactions in the dynamic and stagnant flow regions, it is assumed that the same rate coefficients apply to both soil regions (Ma and Selim, 1997).

In the modified approach, they considered the dynamic and stagnant soil regions in the soil as a continuum, and connected to one another (Figure 8.10). Solute retention may occur concurrently in the dynamic and stagnant regions until equilibrium conditions are attained or all vacant sites for a soil aggregate become occupied (filled). They proposed that the rates of retention reactions in the mobile and immobile phases are a function of the total vacant sites in the soil. Specifically, this modified approach does not distinguish between the fraction of sites associated with the dynamic region and that of the stagnant region. That is, the amount retained from the mobile phase, for example, affects the total number of vacant sites for retention of solutes in the immobile water phase, and vice versa. In fact, the fraction of sites f has been shown to be highly affected by experimental conditions, such as particle size, water flux, solute concentration, and species considered (van Genuchten and Wierenga, 1977; Nkedi-Kizza et al., 1983). Selim and Ma (1995) also assumed that the second-order approach accounts for two reversible kinetic reactions and one irreversible reaction. Specifically, S_e and S_k are associated with reversible and S_i with irreversible reactions. According to the second-order rate law, the rate of reaction is not only a function of solute concentration in solution but also of the number of available retention sites on matrix surfaces. As the sites become filled or occupied by the retained solute, the number of vacant or unfilled sites, which we denote as φ (µg per gram soil), approaches zero. In the mean time, the amount of solutes retained by the soil matrix (S) approaches the total capacity or maximum sorption sites S_{max}.

Incorporating the modified concept into the mobile-immobile approach, the transport convection-dispersion equation with reactions in the dynamic soil region can be rewritten as:

$$\Theta^m \frac{\partial C^m}{\partial t} + \rho \frac{\partial S^m}{\partial t} = \Theta^m D \frac{\partial^2 C^m}{\partial x^2} - v^m \Theta^m \frac{\partial C^m}{\partial x} - \alpha (C^m - C^{im}) \qquad (8.57)$$

and the associated mass transfer equation as:

$$\Theta^{im}\frac{\partial C^{im}}{\partial t}+\rho\frac{\partial S^{im}}{\partial t}=\alpha(C^m-C^{im})\qquad(8.58)$$

Let φ denote the total number of vacant sites. The reactions in the dynamic soil region are:

$$S_e{}^m = K_e\Theta^m C^m\varphi\qquad(8.59)$$

$$\frac{\partial S_k{}^m}{\partial t}=k_3\Theta^m C^m\varphi-k_4S_k{}^m-k_5S_k{}^m\qquad(8.60)$$

$$\frac{\partial S_i{}^m}{\partial t}=k_5S_k{}^m\qquad(8.61)$$

For the stagnant region, the reactions are

$$S_e{}^{im}=K_e\Theta^{im}C^{im}\varphi\qquad(8.62)$$

$$\frac{\partial S_k{}^{im}}{\partial t}=k_3\Theta^{im}C^{im}\varphi-k_4S_k{}^{im}-k_5S_k{}^{im}\qquad(8.63)$$

and

$$\frac{\partial S_i{}^{im}}{\partial t}=k_5S_k{}^{im}\qquad(8.64)$$

Maximum sorption capacity S_{max} for a given soil is related to the total vacant sites φ, according to:

$$S_{max}=\varphi+S_e{}^m+S_k{}^m+S_e{}^{im}+S_k{}^{im}\qquad(8.65)$$

At large reaction time when equilibrium is assumed, the following relations express the amount sorbed for the dynamic and stagnant regions, respectively:

$$\frac{S_e{}^m}{S_{max}}=\left[\frac{K_e\Theta^m C^m}{1+\omega\,\Theta_C^-}\right],\quad\frac{S_k{}^m}{S_{max}}=\left[\frac{K_k\Theta^m C^m}{1+\omega\,\Theta_C^-}\right]\qquad(8.66)$$

$$\frac{S_e{}^{im}}{S_{max}}=\left[\frac{K_e\Theta^{im} C^{im}}{1+\omega\,\Theta_C^-}\right],\quad\frac{S_k{}^{im}}{S_{max}}=\left[\frac{K_k\Theta^{im} C^{im}}{1+\omega\,\Theta_C^-}\right]\qquad(8.67)$$

where ω is similar to that developed in the previous model ($\omega = K_e + K_k$) and represents average concentration in solution, ($\Theta = \Theta^m C^m + \Theta^{im} C^{im}$). The total amount S can thus be expressed by the following simplified (Langmuir) form:

$$\frac{S}{S_{max}} = \frac{\omega \Theta \overline{C}}{1 + \omega \Theta \overline{C}} \tag{8.68}$$

Therefore, the modified mobile-immobile model approaches a one-site Langmuir isotherm. Moreover, the parameter f is absent in Equations 8.41 to 8.52 and, at equilibrium, the amounts sorbed in each fraction depend solely on K_e, K_k, and Θ^m. Therefore, the partitioning coefficient f of the two-region concept, which is a difficult-to-measure parameter, need not be specified and the amount retained by each soil region is solely a function of the rates of reactions.

Selim and Ma (1995) applied their modified model on miscible displacement results for atrazine. They showed that their modified approach provided good BTC predictions for a wide range of aggregate sizes and flow velocities for a Sharkey clay soil. They also concluded that the modified approach, which requires fewer parameters, is superior to the original model of Selim and Amacher (1988). Measured and predicted atrazine BTCs from a soil column with 4 to 6 mm aggregated Sharkey soil are given Figure 8.18. Predictions shown are based on the second-order mobile-immobile approach with different values of the fraction of mobile water f. In Figure 8.19, measured and predicted BTCs for two columns of Sharkey clay soil having two aggregate sizes; 2 to 4 mm and 4 to 6 mm in diameter. In this example, the second-order model (SOM) is compared with the coupled second-order model and the immobile-immobile approach (SOM-MIM). The predictions are based on two estimates

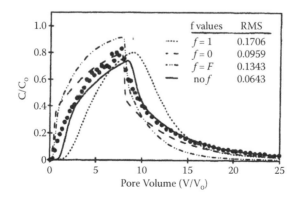

FIGURE 8.18
Measured and predicted atrazine BTC for 4- to 6-mm aggregated Sharkey. Continuous and dashed curves are simulations using the second-order mobile-immobile approach with different values of the fraction of mobile water f.

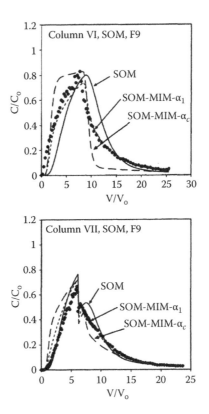

FIGURE 8.19

Measured and predicted atrazine BTCs for two columns of Sharkey clay soil having two aggregate sizes. Top figure is for a 10-cm column packed with 4- to 6-mm soil aggregates with no flow interruption, and bottom figure is for a 15-cm column packed with 2- to 4-mm soil aggregates, 6 d flow interruption after pulse input. Predictions are based on the second-order mobile-immobile model (MIM-SOM) and the original second-order model (SOM).

of the transfer rate coefficient (α) using Equations 8.14 and 8.16. These examples illustrate the capability of SOM-MIM in describing the BTC as well as the influence of flow interruption. Based on these predictions, this modified approach is a promising one since it accounts for retention and transport processes based on physically as well as chemically heterogeneous reactions.

Figure 8.20 presents the BTCs from column experiments on arsenite transport in Olivier loam, which displays diffusive fronts followed by extensive tailing or slow release during leaching. Sharp decrease or increase in arsenite concentration after flow interruption further verified the extensive nonequilibrium condition. The arsenite BTC was simulated using coupled physical and chemical nonequilibrium approaches. Here physical transport

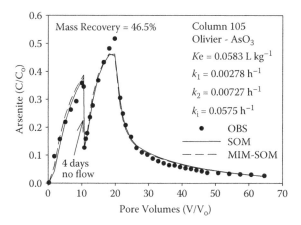

FIGURE 8.20
Measured and predicted arsenite for Olivier soil column. Predictions are based on the second-order mobile-immobile model (MIM-SOM) and the original second-order model (SOM).

parameters (D, f, and α) were based on the estimated values from tritium BTCs, whereas kinetic retention parameters (K_e, k_1, k_2, k_i) were obtained from inverse modeling. The incorporation of the physical nonequilibrium approach resulted in a slightly sharper front, reflecting the effect of preferential flow path. The coupled model successfully predicted the concentration drop during flow interruption on the adsorption side. Based on the results of model predictions, we concluded that kinetics of arsenite reaction on soil surface is the dominant process of arsenite transport in soils. The utilization of the physical nonequilibrium parameters in the combined model is expected to increase for modeling contaminant transport in heterogeneous natural porous media compared with columns repacked in the laboratory with homogenized soils.

Although the two-region model concept has been shown to successfully describe the appearance of lack of equilibrium behavior and tailing for a wide range of conditions, this approach has several drawbacks. First, the value of α is difficult to determine for soils because of the irregular geometric distribution of immobile water pockets, and second, the fraction of mobile and immobile water within the system can only be estimated. Thus, two parameters are needed (for nonreacting solute) and they can only be found by curve fitting of the flow equations to effluent data. Another drawback of the mobile-immobile approach is the inability to identify unique retention reactions associated with the dynamic and stagnant soil regions separately. Due to this difficulty, a general assumption implicitly made is that similar processes and associated parameters occur within both regions. Thus, a common set of model parameters is utilized for both regions. Such an assumption has been made for equilibrium (linear, nonlinear, and ion exchange) as well as kinetic reversible and irreversible reactions. Therefore, this model disregards the heterogeneous nature of various types of sites on matrix surfaces. This is not surprising since soils are not homogeneous systems but, rather, are a complex mixture of solids of clay minerals, several oxides/hydroxides, and organic matter with varying surface properties.

8.6 Multiflow Domain Approaches

Following the basic structure of the mobile-immobile approach, other approaches were developed to describe the observed deviation of nonreactive solute transport based on the classical convection-dispersion equation for one-region or -domain. Common among such models is the assumption that the soil consists of several flow regions or domains where each flow domain is characterized by different a different water flux, soil water content and dispersion coefficients. This is in contrast with the mobile-immobile approach where a stagnant (no-flow) domain or region is assumed. Without loss of generality, a convective-dispersion flow equation for nonreactive solutes in domain i may be expressed as:

$$\frac{\partial C_i}{\partial t} = D_i \frac{\partial^2 C_i}{\partial x^2} - v_i \frac{\partial C_i}{\partial x} \tag{8.69}$$

where D_i, Θ_i, and v_i are the dispersion coefficient, soil water content, and pore-water velocity associated with each domain I, where

$$\Theta = \sum_{i=1}^{l} \Theta_i \quad \text{and} \quad q = \sum_{i=1}^{l} q_i \tag{8.70}$$

and the pore-water velocity v for domain i is

$$v_i = \frac{q_i}{\Theta_i} \tag{8.71}$$

The two-domain transport model is the simplest case where one divides soil water into two regions based on their flow velocities. Both water regions have a non-zero flow rate. Without loss of generality, we denoted the fast flow region as A and slow flow region as B. The soil system was characterized by velocity (q_A, q_B), porosity (Θ_A, Θ_B), solute concentration (C_A, C_B), and dispersion coefficient (D_A, D_B). The two flow domains are related by an interaction term Γ such that:

$$\Gamma = \alpha(C_A - C_B) \tag{8.72}$$

where α is the mass transfer coefficient between the two flow domains. This two-domain model reduces to a capillary bundle model when $\alpha = 0$ and it approaches the classic one-domain convective-dispersion equation as α increases. The two-flow domain model can also be reduced to a mobile-immobile model when V_B equals zero. The convective-dispersion equation in each flow domain can be written as (Skopp, Gardner, and Tyler, 1981):

$$\frac{\partial C_A}{\partial t} = D_A \frac{\partial^2 C_A}{\partial x^2} - v_A \frac{\partial C_A}{\partial x} - \frac{\alpha}{\Theta_A}(C_A - C_B) \tag{8.73}$$

$$\frac{\partial C_B}{\partial t} = D_B \frac{\partial^2 C_B}{\partial x^2} - v_B \frac{\partial C_B}{\partial x} - \frac{\alpha}{\Theta_B}(C_B - C_A) \tag{8.74}$$

The corresponding initial and boundary conditions associated with each flow domain (i) can be expressed as:

$$C_i = 0 \quad (t = 0, 0 < x < L) \tag{8.75}$$

$$v_i C_o = vC_i - \Theta_i D_i \frac{\partial C_i}{\partial x} \qquad (x = 0, t < t_p) \tag{8.76}$$

$$0 = vC_i - \Theta_i D_i \frac{\partial C_i}{\partial x} \qquad (x = 0, t > t_p) \tag{8.77}$$

$$\frac{\partial C_i}{\partial x} = 0 \quad (x = L, t \geq 0) \tag{8.78}$$

These conditions are similar to those described earlier for the transport of a solute pulse (input) having a concentration C_o in a uniform soil having a finite length L where a steady water flux v was maintained constant.

8.6.1 Experimental Evidence

The two-flow domain model, which may be referred to as a dual-porosity concept, has been used in the solute transport literature to account for the contribution of large pores to preferential flow. Jarvis, Bergstrom, and Dik (1991) applied the dual porosity model to a field study of water-unsaturated and transient flow conditions. They found that a two-flow domain model provided improved description of chloride movement compared with a one-flow domain model. Gerke and van Genuchten (1993) also applied a dual-porosity model to simulate the preferential movement of water in structured porous media. Othmer, Diekkruger, and Kutilek (1991) found that the hydraulic conductivity $[K(\psi)]$ derived from a bimodal pore size distribution was in agreement with the field-measured $K(\psi)$. Here the term ψ refers to the water suction (cm). Othmer, Diekkruger, and Kutilek (1991) concluded that the bimodal porosity model represented the reality of soil system better than a unimodal porosity model. Similar results were reported by Smettem and Kirkby (1990) in an aggregate soil. Wilson, Jardine, and Gwo (1992) presented a three-flow domain model to describe the $K(\psi) - \psi$ or $K(\Theta) - \Theta$ relationship in different soil horizons.

Delineation of a flow domain in a multidomain approach has been at best vague and somewhat arbitrary. Commonly used concepts rely on the definition of macropores and micropores, which are also arbitrary. The only known experimental method for estimating two porosities is based on a measure of the soil water content at an arbitrarily chosen water tension (ψ). From the shape of the soil-moisture characteristic curve, Smettem and Kirkby (1990) chose ψ of 14 cm as the matching point between the interaggregate (macro-) and the intraaggregate (micro-) porosity. Jarvis, Bergstrom, and Dik (1991) estimated macroporosity based on measured specific yield under a water tension of 100 cm. Other water tensions used to differentiate macropores from micropores include 10 cm by Wilson, Jardine, and Gwo (1992), 20 cm by Selim, Schulin, and Flühler (1987), and 80 cm by Nkedi-Kizza et al. (1982). Field and laboratory observations that support the two-flow domain concept are the bimodal peaks for nonreactive solutes when steady soil water flow was dominant. Hamlen and Kachanoski (1992) observed bimodal chloride breakthrough in the B horizon of a sandy soil. However, the bimodal peaks were not observed in the A horizon. Skopp, Gardner, and Tyler (1981) suspected that solute bimodal peaks were caused by the presence of dual porosities in the soil.

The examples presented here to illustrate physical two-flow domains or dual porosity are those of the work of Ma and Selim (1995), who investigated transport of the nonreactive tracer tritium in packed columns. Ma and Selim (1995) carried out a series of experiments to examine the effect of water flux on the shape of the BTCs. Figure 8.21 shows BTCs for three Mahan soil columns having flow velocities of 2.02, 3.82, and 5.28 cm/h, respectively. These BTCs are for tritium pulse applications of approximately one pore volume. From Figure 8.21, tailing of the desorption side was observed for all three BTCs. A hump was also observed on the

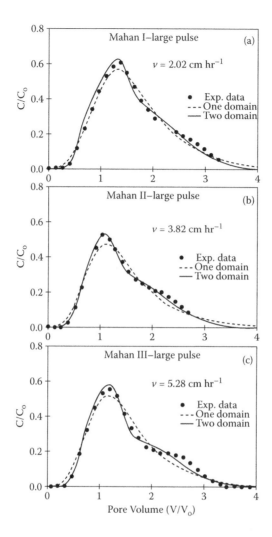

FIGURE 8.21
Tritium breakthrough curves (BTCs) for three Mahan soil columns under different fluxes. Solid and dashed curves are fitted BTCs with the one- and two-flow domain models, respectively.

tailing side, which became more obvious as the velocity increased. These observed humps or peaks suggest a nonuniform flow domain in such a soil. The humps or second peaks were more distinct for the tritium BTCs shown in Figure 8.22. In fact, the peaks developed into a bimodal peak at the highest flow velocity (bottom of Figure 8.22). All three soil columns shown in Figure 8.22 received a small pulse of tritium of 0.02 pore volumes. Based on the BTCs shown in Figure 8.22, it is apparent that the

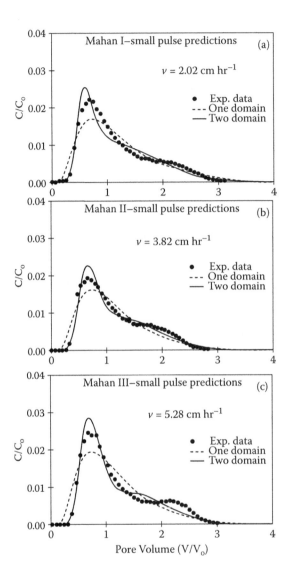

FIGURE 8.22
Tritium breakthrough curves (BTCs) for three Mahan soil columns under different fluxes. The applied tritium pulse was 0.02 pore volume. Solid and dashed curves are fitted BTCs with the one- and two-flow domain models, respectively.

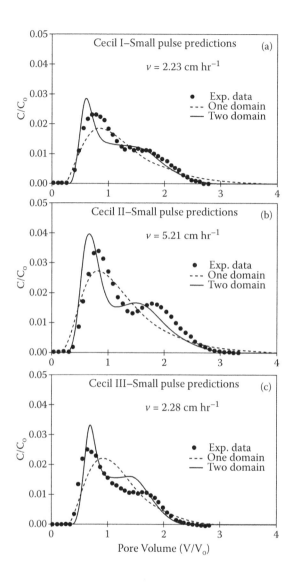

FIGURE 8.23
Tritium breakthrough curves (BTCs) for three Cecil soil columns under different fluxes. The applied tritium pulse was 0.02 pore volume. Solid and dashed curves are fitted BTCs with the one- and two-flow domain models, respectively.

shape of the BTC depended on pulse duration and flow velocity. Those columns that received a small tritium pulse also showed early breakthrough, with the first peak position well within one pore volume. Similar BTC results were obtained for Cecil and Dothan soils, as shown in Figures 8.23 and 8.24, respectively.

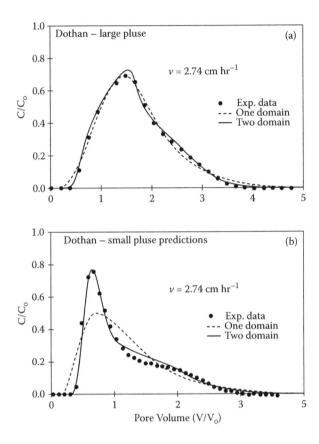

FIGURE 8.24
Tritium breakthrough curves (BTCs) for two Dothan soil columns with applied tritium pulses of one pore volume (top figure) and 0.02 pore volume (bottom figure). Solid and dashed curves are fitted BTCs with the one- and two-flow domain models, respectively.

The tritium BTCs presented in Figures 8.21 to 8.24 were fitted using both the one- and two-flow domain models and are shown by the solid and dashed curves. Although the two-flow domain model was superior over the one-domain model in tritium BTC description, the BTCs from the large pulse (one pore volume) were not perfectly described, especially on the tailing side (Figures 8.21–8.24). Improvement in BTC fitting is difficult since independent estimates of α based on experimental methods to differentiate between macropores and micropores are not available. Nevertheless, the two-flow domain was capable of capturing the double peaks of tritium for three different soils.

References

Barber, S. A. 1984. *Soil Nutrient Bioavailability: A Mechanistic Approach*. New York: Wiley Interscience.

Biggar, J. W., and D. R. Nielsen. 1976. Spatial variability of leaching characteristics of a field soil. *Water Resour. Res.* 12: 78–84.

Brenner, H. 1962. The diffusion model of longitudinal mixing in beds of finite length, Numerical values. *Chem. Eng. Sci.* 17: 220–243.

Brusseau, M. L. 1993. The influence of solute size, pore water velocity, and intraparticle porosity on solute-dispersion and transport in soil. *Water Resour. Res.* 29: 1071–1080.

Brusseau, M. L., P. S. C. Rao, R. E. Jessup, and J. M. Davidson. 1989. Flow interruption: A method for investigating sorption nonequilibrium. *J. Contam. Hydrol.* 4: 223–240.

Buchter, B., C. Hinz, M. Flury, and H. Flühler. 1995. Heterogeneous flow and solute transport in an unsaturated stony soil monolith. *Soil Sci. Soc Am. J.* 59: 14–21.

Cameron, D., and A. Klute. 1977. Convective-dispersive solute transport with chemical equilibrium and kinetic adsorption model. *Water Resour. Res.* 13: 183–188.

Camobreco, V. J., B. K. Richards, T. S. Steenhuis, J. H. Peverly, and M. B. McBride. 1996. Movement of heavy metals through undisturbed and homogenized soil columns. *Soil Sci.* 161: 740–750.

Chen, C. and R. J. Wagenet. 1992. Simulation of water and chemicals in macropore soils. I. Representation of the equivalent macropore influence and its effect on soil water flow. *J. Hydrol.* 130: 105–126.

Clothier, B. E., M. B. Kirkham, and J. E. Mclean. 1992. In situ measurement of the effective transport volume for solute moving through soil. *Soil Sci. Soc. Am. J.* 56: 733–736.

Coats, K. H., and B. D. Smith. 1964. Dead-end pore volume and dispersion in porous media. *Soc. Pet. Eng.* 4: 73–84.

Corwin, D. L., A. David, and S. Goldberg. 1999. Mobility of arsenic in soil from the Rocky Mountain Arsenal area. *J. Contam. Hydrol.* 39: 35–58.

Dagan, G. 1984. Solute transport in heterogeneous porous formations, *J. Fluid Mech.* 145: 151–177.

De Smedt, F., and P. J. Wierenga. 1984. Solute transfer through columns of glass beads. *Water Resour. Res.* 20: 225–232.

Goltz, M. N., and P. V. Roberts. 1988. Simulations of physical nonequilibrium solute transport models: application to a large scale field experiment. *J. Contamin. Hydrol.* 3: 37–63.

Gaston, L. A., and H. M. Selim. 1990. Transport of exchangeable cations in an aggregated clay soil. *Soil Sci. Soc. Am. J.* 54: 31–38.

Gaston, L. A., and H. M. Selim. 1990a. Transport of exchangeable cations in an aggregated clay soil. *Soil Sci. Soc. Am. J.* 54: 31–38.

Gaston, L. A., and H. M. Selim. 1990b. Prediction of cation mobility in montmorillonitic media based on exchange selectivities of montmorillonite. *Soil Sci. Soc. Am. J.* 54: 1525–1530.

Gerke, H. H., and M. Th. van Genuchten. 1993. Evaluation of a first-order water transport term for variably saturated dual-porosity flow models. *Water Resour. Res.* 29: 1225–1238.

Han, F. X., W. L. Kingery, H. M. Selim, and P. D. Gerard. 2000. Accumulation of heavy metals in a long term poultry waste amended soil. *Soil Sci.* 165: 260–268.

Hamlen, C. J., and R. G. Kachanoski. 1992. Field solute transport across a soil horizon boundary. *Soil Sci. Soc. Am. J.* 56: 1716–1720.

Hu, Q., and M. L. Brusseau. 1995. The effect of solute size on transport in structured porous media. *Water Resour. Res.* 31: 1637–1646.

Jarvis, N. J., L. Bergstrom, and P. E. Dik. 1991. Modeling water and solute transport in macroporous soil. II. Chloride breakthrough under non-steady flow. *J. Soil Sci.* 42: 71–81.

Jarvis, N. J., K. G., Villholth and B. Ule´n 1999. Modelling particles mobilization and leaching in macroporous soil. *Eur. J. Soil Sci.* 50: 621–632.

Jaynes, D. B., R. S. Bowman, and R. C. Rice. 1988. Transport of a conservative tracer in a field under continuous flood irrigation. *Soil Sci. Soc. Am. J.* 52: 618–624.

Jaynes, D. B., S. D. Logsdon, and R. Horton. 1995. Field method for measuring mobile-immobile water content and solute transfer rate coefficient. *Soil. Sci. Soc. Am. J.* 59: 352–356.

Jury, W. A., and H. Flühler. 1992. Transport of chemicals through soil: Mechanisms, models, and field applications. *Adv. Agron.* 47: 141–201.

Koch, S., and H. Flühler. 1993. Non-reactive transport with micropore diffusion in aggregated porous media determined by a flow interruption technique. *J. Contam. Hydrol.* 14: 39–54.

Li, Y. and M. Ghodrati. 1994. Preferential transport of nitrate through soil columns containing root channels. *Soil Sci. Soc.Am. J.* 58: 653–659.

Luxmoore, R. J. 1981. Micro-, meso-, and microporosity of soil. *Soil Sci. Soc. Am. J.* 45: 671.

Ma, L., and H. M. Selim. 1995. Transport of nonreactive solutes in soils: A two-flow domain approach. *Soil Sci.* 159: 224–234.

Ma, L. and H. M. Selim. 1997. Evaluation of nonequilibrium models for predicting atrazine transport in soils. *Soil Sci. Soc. Am. J.* 1299–1307.

Murali, V., and A.G. Aylmore. 1980. No-flow equilibrium and adsorption dynamics during ionic transport in soils. *Nature (London)* 283: 467–469.

Nkedi-Kizza, P., J. M. Biggar, H. M. Selim, M. Th. van Genuchten, P. J. Wierenga, J. M. Davidson, and D. R. Nielsen. 1983. Modeling tritium and chloride 36 transport through aggregated oxisols. *Water Resour. Res.* 20: 691–700.

Nkedi-Kizza, P., J. M. Biggar, H. M. Selim, M. Th. van Genuchten, P. J. Wierenga, J. M. Davidson, and D. R. Nielsen. 1984. On the equivalence of two conceptual models for describing ion exchange during transport through an aggregated soil. *Water Resour. Res.* 20: 1123–1130.

Nkedi-Kizza, P., P. S. C. Rao, R. E. Jessup, and J. M. Davidson. 1982. Ion exchange and diffusive mass transfer during miscible displacement through an aggregated Oxisol. *Soil Sci. Soc. Am. J.* 46: 471–476.

Othmer, H., B. Diekkruger, and M. Kutilek. 1991. Bimodal porosity and unsaturated hydraulic conductivity. *Soil Sci.* 152: 139–150.

Parker, J. C., and M. Th. van Genuchten. 1984. Determining transport parameters from laboratory and field tracer experiments. *Virginia Agric. Exp. Stn. Bull.* 84-3.

Parker, J. C., and A. J. Valocchi. 1986. Constraints on the validity of equilibrium and first-order kinetic transport model in structured soils. *Water Resour. Res.* 22: 399–407.

Passioura, J. B. 1971. Hydrodynamic dispersion in aggregated media. I. Theory. *Soil Sci.* 111: 339–344.

Rao, P. S. C., J. M. Davidson, R. E. Jessup, and H. M. Selim. 1979. Evaluation of conceptual models for describing nonequilibrium adsorption-desorption of pesticide during steady-state flow in soils. *Soil Sci. Soc. Am. J.* 43: 22–28.

Rao, P. S. C., R. E. Jessup, D. E. Ralston, J. M. Davidson, and D. P. Kilcrease. 1980a. Experimental and mathematical description of nonadsorbed solute transfer by diffusion in spherical aggregate. *Soil Sci. Soc. Am. J.* 44: 684–688.

Rao, P. S. C., D. E. Ralston, R. E. Jessup, and J. M. Davidson. 1980b. Solute transport in aggregated porous media: Theoretical and experimental evaluation. *Soil Sci. Soc. Am. J.* 44: 1139–1146.

Reedy, O. C., P. M. Jardine, G. V. Wilson, and H. M. Selim. 1996. Quantifying the diffusive mass transfer of nonreactive solutes in columns of fractured saprolite using flow interruption. *Soil Sci. Soc. Am. J.* 60: 137–384.

Rubin, J. 1983. Transport of reactive solutes in porous media: Relation between mathematical nature of problem formulation and chemical nature of reactions. *Water Resour. Res.* 19: 1231–1252.

Schulin, R., P. J. Wierenga, H. Flühler, and J. Levenberga. 1987. Solute transport through a stony soil. *Soil Sci. Soc. Am. J.* 51: 342.

Selim, H. M., J. M. Davidson, and R. S. Mansell. 1976. Evaluation of a two-site adsorption-desorption-model for describing solute transport in soils. In: Proceedings of Summer Computer Simulation Conference, Washington, D.C., Simulation Councils, La Jolla, CA, p. 444–448.

Selim, H. M., and L. Ma. 1995. Transport of reactive solutes in soils: a modified two-region approach. *Soil Sci. Soc. Am. J.* 59: 75–82.

Selim, H. M. 1992. Modeling the transport and retention of inorganics in soils. *Adv. Agron.* 47: 331–384.

Selim, H. M., and M. C. Amacher. 1988. A second-order kinetic approach for modeling solute retention and transport in soils. *Water Resources Res.* 24: 2061–2075.

Selim, H. M., and M. C. Amacher. 1997. *Reactivity and Transport of Heavy Metals in Soils.* Boca Raton, FL: CRC Press.

Selim, H. M., R. Schulin, and H. Flühler. 1987. Transport and ion exchange of calcium and magnesium in an aggregated soil. *Soil Sci. Soc. Am. J.* 51: 876–884.

Simunek, J., N. J. Jarvis, M. T. van Genuchten, and A. Gardenas. 2003. Review and comparison of models for describing non-equilibrium and preferential flow and transport in the vadose zone. *J. Hydrol.* 272: 14–35.

Skopp, J., W. R. Gardner, and E. J. Tyler. 1981. Solute movement in structured soils: Two-region model with small interaction. *Soil Sci. Soc. Am. J.* 45: 837–842.

Smettem, K. R. J. 1984. Soil-water residence time and solute uptake. III. Mass transfer under simulated winter rainfall conditions in undisturbed soil cores. *J. Hydrol. (Amsterdam)* 67: 235–248.

Smetten K. R. J., and C. Kirby. 1990. Measuring the hydraulic properties of a stale aggregated soil. *J. Hydrol.* 117: 1–13.

Sterckeman, T., F. Douay, N, Proix, and H. Fourrier. 2000. Vertical distribution of Cd, Pb and Zn in soils near smelters in the North of France. *Environ. Pollut.* 107: 377–389.

Valocchi, A. J. 1985. Validity of the local equilibrium assumption for modeling sorbing solute transport through homogeneous soils. *Water Resour. Res.* 21: 808–820.

van Eijkeren, J. C. M., and I. P. G. Lock. 1984. Transport of cation solutes in sorbing porous medium. *Water Resour. Res.* 20: 714–718.

van Genuchten, M. Th. 1981. Non-equilibrium transport parameters from miscible displacement experiments. Research Report No. 119. Riverside, CA: U. S. Salinity Laboratory.

van Genuchten, M. Th., and F. N. Dalton. 1986. Models for simulating salt movement in aggregated field soils. *Geoderma* 38: 165–183.

van Genuchten, M. Th., and P. J. Wierenga. 1976. Mass transfer studies in sorbing porous media. I. Analytical solutions. *Soil Sci. Soc. Am. J.* 40: 473–480.

van Genuchten, M. Th., and P. J. Wierenga. 1977. Mass transfer studies in sorbing porous media. II. Experimental evaluation with tritium (^3H_2O). *Soil Sci. Soc. Am. J.* 41: 272–277.

van Genuchten, M. Th., P. J. Wierenga, and G. A. O'Connor. 1977. Mass transfer studies in sorbing porous media. III. Experimental evaluation of 2.4.5-T. *Soil Sci. Soc. Am. J.* 41: 278–284.

van Genuchten. 1985. Convective-dispersive transport of solutesin sequential first decay reactions. Comp. Geosci. 11: 129–147.

van Genuchten, M. Th. and R. J. Wagenet. 1989. Two-site/two-region models for pesticide transport and degradation: Theoretical development and analytical solutions. *Soil Sci. Soc. Am. J.* 53: 1303–1310.

Villermaux, J. 1974. Deformation of chromatographic peaks under the influence of mass transfer phenomena. *J. Chromatographic Sci.* 12: 822–831.

Wilson, G. V., P. M. Jardine, and J. P. Gwo. 1992. Modeling the hydraulic properties of a multiregion soil. *Soil Sci. Soc. Am. J.* 56: 1731–1737.

9

Transport in Nonuniform Media

Soils are defined as heterogeneous systems that are made up of various constituents having distinct physical and chemical properties. A major source of soil heterogeneity is soil stratifications or layering. The phenomenon of stratifications or layering of the soil profile is an intrinsic part of soil formation processes and has been documented for several decades by soil survey work.

Another source of soil heterogeneity is due to mixing of different soils or geological media. Mixing of soils and/or geological media may be the result of various industrial and agricultural activities. Such activities include site remediation, construction and land leveling, incorporation of industrial and municipal waste, manure application and mixing, and crop residue management. Therefore, approaches that account for the reactivities in mixed media are necessary to predict the fate and transport of reactive dissolved chemicals in such media.

9.1 Layered Media

The transport processes of dissolved chemicals in stratified or layered soils have been studied for several decades by Shamir and Harleman (1967), Selim, Davidson, and Rao (1977), Bosma and van der Zee (1992), and Wu, Kool, and Huyakorn (1997), among others. Solute transport in layered soils can be investigated through numerical methods as well as approximate analytical solutions. An early analytical method was proposed by Shamir and Harleman (1967), who used a system's analysis approach. They assumed that different layers were independent with regard to solute travel time. Each layer's response served as the boundary condition for the downstream layer and so on. Later, Selim, Davidson, and Rao (1977) discussed the movement of reactive solutes through layered soils using finite-difference numerical methods. They considered both equilibrium and kinetic sorption models of the linear and nonlinear types. In the late 1980s, Leij and Dane (1989) developed analytical solutions for the linear sorption-type models using Laplace transforms. Their solutions were based on the assumption that each layer was semi-infinite. Bosma and van der Zee (1992) also proposed an approximate analytical solution for reactive solute transport in layered soils using an adaptation of the traveling wave solution. Wu, Kool, and Huyakorn (1997) developed another analytical model for nonlinear adsorptive transport through layered soils ignoring the effects of dispersion. In addition, Guo et al. (1997) showed that the transfer function approach was a very powerful tool to describe the nonequilibrium transport of reactive solutes through layered soil profiles with depth-dependent adsorption.

When we study transport processes of dissolved chemicals in layered soils, it is of interest to investigate whether soil layering affects solute breakthrough. When flow remains one-dimensionally vertical, which is the case when horizontal stratification is dominant, it is of interest whether the layering order affects breakthrough results at the groundwater level (van der Zee, 1994). The early results from Shamir and Harleman (1967) showed that the order of layering did not affect breakthrough significantly. This interesting result was further elaborated upon by Barry and Parker (1987) based on various analytical approaches. Results from various linear and nonlinear numerical simulations for several sorption model types also supported this conclusion (Selim, Davidson, and Rao, 1977). Furthermore, Selim, Davidson, and Rao (1977) concluded that layering order was also unimportant for Freundlich adsorption. Their experimental results also supported this conclusion. However, van der Zee (1994) attributed the results attained by Selim and coworkers (1977) to the small Peclet number assumed for the nonlinear layer, which prevents nonlinearity effects from being clearly manifested. Van der Zee (1994) used a hypothetical result to illustrate that layering sequence should have an effect. However, what van der Zee (1994) used to support his conclusion was the traveling wave, which was the curve of concentration versus depth at different times, that is, the concentration profile. More recently, Zhou and Selim (2001) accounted for several nonlinear and kinetic retention mechanisms for multilayered soils. For individual soil layers, Zhou and Selim (2001) considered solute retention mechanisms of the nonlinear (Freundlich), Langmuir, first- and nth-order kinetic, second-order kinetic, and irreversible reactions. For all retention mechanisms used, their simulation results indicated that solute breakthrough curves (BTCs) were similar regardless of the layering sequence in a soil profile. This finding is consistent with the earlier finding of Selim, Davidson, and Rao (1977) and contrary to that of Bosma and van der Zee (1992) for nonlinear adsorption.

In the next sections, we present general equations for solute transport in multilayered systems followed by a discussion of the various choices of boundary conditions at the interface between layers. We further present selected case studies of linear and nonlinear adsorption mechanisms in layered soil systems based on numerical simulations for a range of soil properties and fluxes (or Brenner numbers). Experimental BTCs based on miscible displacements from a packed sand-clay soil column for a tracer (tritium) and for reactive solutes (Ca and Mg) are subsequently presented in support of theoretical findings.

9.2 Convection-Dispersion Equation

A two-layered soil column of length L is shown in Figure 9.1. The length of each layer is denoted by L_1 and L_2, respectively. To show heterogeneity, each soil layer has specific, but not necessarily the same, water content, bulk density, and solute retention properties. Only vertical, steady-state water flow perpendicular to the soil layers (Figure 9.1) will be considered. The convective-dispersive equation (CDE) governing solute transport in the ith layer (see Figure 9.1) is given by Equation 9.1 (Selim, Davidson, and Rao 1977):

$$\rho_i \frac{\partial S_i}{\partial t} + \theta_i \frac{\partial C_i}{\partial t} = \frac{\partial}{\partial x} \theta_i D_i \frac{\partial C_i}{\partial x} - q \frac{\partial C_i}{\partial x} - Q_i$$

(9.1)

$$(0 \le x \le L_i, i = 1, 2)$$

where (omitting the i):

C = resident concentration of solute in soil solution ($\mu g\ cm^{-3}$)
S = amount of solute adsorbed by the soil matrix ($\mu g\ g^{-1}$)
ρ = soil bulk density ($g\ cm^{-3}$)
θ = volumetric soil water content ($cm^3\ cm^{-3}$)
D = solute dispersion coefficient ($cm^2\ d^{-1}$)
q = Darcy soil-water flow velocity ($cm\ d^{-1}$)
Q = a sink or source for irreversible solute interaction ($\mu g\ cm^{-3}\ d^{-1}$)
x = distance from the soil surface (cm)
t = time (d)

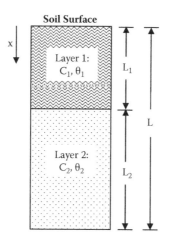

FIGURE 9.1
A schematic diagram of a two-layered soil.

The reversible solute retention from the soil solution is represented by the term, on the left side of Equation 9.1, while the irreversible solute removed from soil solution is expressed by the term Q on the right side of Equation 9.1.

9.3 Boundary Condition at the Interface

An important boundary condition needed in the analysis of multilayered soils is that at the interface between layers. It should be noted that both first-type and third-type boundary conditions are applicable at the interface. Leij, Dane, and van Genuchten (1991) showed that although the principle of solute mass conservation is satisfied, a discontinuity in concentration develops when a third-type interface condition is used. On the other hand, a first-type interface condition will result in a continuous concentration profile across the boundary interface at the expense of solute mass balance. To overcome the limitations of both first- and third-type conditions, a combination of first- and third-type conditions was implemented. The first-type condition can be written as:

$$C_I\big|_{x \to L_1^-} = C_{II}\big|_{x \to L_1^+}, \quad t > 0 \tag{9.2}$$

where and denote that $x = L_1$ is approached from the upper and lower layer, respectively. Similarly, the third-type condition can be written as:

$$\left(qC_I - \theta_I D_I \frac{\partial C_I}{\partial x} \right)\Bigg|_{x \to L_1^-} = \left(qC_{II} - \theta_{II} D_{II} \frac{\partial C_{II}}{\partial x} \right)\Bigg|_{x \to L_1^+}, \quad t > 0 \tag{9.3}$$

Incorporation of Equation 9.2 into Equation 9.3 yields:

$$\theta_I D_I \frac{\partial C_I}{\partial x}\Bigg|_{x \to L_1^-} = \theta_{II} D_{II} \frac{\partial C_{II}}{\partial x}\Bigg|_{x \to L_1^+}, \quad t > 0 \tag{9.4}$$

The boundary condition of Equation 9.4 was first proposed by Zhou and Selim (2001) and resembles that for a second-type boundary condition as indicated earlier by Leij, Dane, and van Genuchten (1991).

9.4 Equilibrium Retention Models

The form of solute retention reactions in the soil system must be identified if prediction of the fate of reactive solutes in the soil using the CDE (Equation 9.1) is sought. The reversible term $(\partial S/\partial t)$ is often used to describe the rate of sorption or exchange reactions with the solid matrix. Sorption or exchange has been described by either instantaneous equilibrium or a kinetic reaction where concentrations in solution and sorbed phases vary with time. Linear, Freundlich, and one- and two-site Langmuir equations are perhaps the most commonly used to describe equilibrium reactions. In the subsequent sections we discuss Freundlich and Langmuir reactions and their use in describing equilibrium retention. This is followed by kinetic-type reactions and their implication for single and multireaction retention and transport models.

9.4.1 Linear Adsorption

Figure 9.2 shows a comparison of breakthrough curves (BTCs) for a two-layered soil column with reverse layering orders. Here we report results for a layered soil column where one layer is nonreactive ($R = 1$) and the other is linearly adsorptive. For the case of linear adsorption, a dimensionless retardation factor can be obtained from Equation 9.1 and is given by:

$$R = 1 + \frac{\rho K_d}{\theta} \tag{9.5}$$

where K_d is a redistribution coefficient where the linear equilibrium model was assumed:

$$S = K_d C \tag{9.6}$$

For the case where $K_d = 0$, the retardation factor R equals 1 and solute is considered nonreactive. The BTC for the case R1 $\rightarrow$ R2 where the nonreactive layer was first encountered (top layer) was similar to that when the

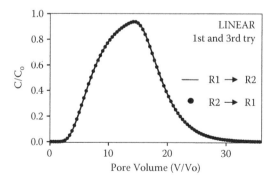

FIGURE 9.2
Simulated breakthrough results for a two-layered soil column under different layering orders (R1 $\rightarrow$ R2 and R2 $\rightarrow$ R1). Here R1 is a nonreactive layer and R2 is a reactive layer with linear adsorption.

layering sequence was reversed (R2 → R1) and the reactive layer (R2) was the top layer. Therefore, for the linear adsorption case, one concludes that the order of soil stratification or layering sequence fails to influence solute BTCs, consistent with those reported earlier by Shamir and Harleman (1967) and Selim, Davidson, and Rao (1977) for systems with two or more layers. Based on these results, a layered soil profile could be regarded as homogeneous with an average retardation factor used to calculate effluent concentration distributions. An average retardation factor R for N-layered soil can simply be obtained from:

$$\overline{R} = \frac{1}{L} \sum_{i=1}^{N} R_i L_i \qquad (9.7)$$

BTCs identical to those in Figure 9.2 were obtained using the solution to the convection-dispersion Equation 9.7 presented by van Genuchten and Alves (1982) and an average retardation factor. This averaging procedure (Equation 9.7) can also be used to describe the BTCs from a soil profile composed of three or more layers. However, if solute distribution within the profile is desired, the use of an average retardation factor is no longer valid and the problem must be treated as a multilayered case.

9.4.2 Nonlinear Freundlich Adsorption

Simulated BTCs of solutes from a two-layered soil system with one as a nonlinear (Freundlich) adsorptive layer are given in Figure 9.3. K_d is the

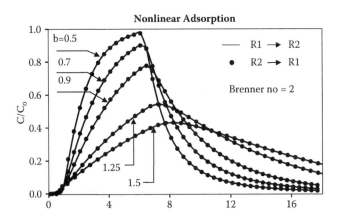

FIGURE 9.3
Simulated breakthrough results for a two-layered soil column under different layering orders (R1 → R2 and R2 → R1). Here R1 is a nonreactive layer and R2 is a reactive layer with nonlinear adsorption, with b values (Equation 9.11) of 0.5, 0.7, 0.9, 1.25, and 1.5. The Brenner number B used was 2.

redistribution coefficient where the linear equilibrium model was assumed:

$$S = K_d C^b \tag{9.8}$$

where the exponent b is commonly less than unity for most reactive chemicals. Here, R1 represents a nonreactive layer and R2 represents a nonlinear (Freundlich) adsorptive layer. These simulations were carried out for a wide range of the Peclet or Brenner number ($B = qL/\theta D$). We also examined the influence of the nonlinear Freundlich parameter b on the shape of the BTCs. For most reactive chemicals, including pesticides and trace elements, b is always less than unity (Selim and Amacher, 1997). Based on these simulations, BTCs were not influenced by the layering sequences regardless of the Brenner number B when nonlinear Freundlich adsorption was considered. This result is similar to that of Selim, Davidson, and Rao (1977). Dispersion is dominant for the case where the Brenner number is small, whereas convection becomes the dominant process for large B values. The BTCs exhibit increasing retardation or delayed arrival, and excessive tailing of the right-hand side of the BTCs for increasing values of the nonlinear adsorption parameter b. In addition, the BTCs become less spread (i.e., a sharp front) with increasing Brenner numbers. All such cases provide similar observations, that is, the effects of nonlinearity of adsorption are clearly manifested. Nevertheless, for all combinations of b and the Brenner number B used in these simulations, the BTCs under reverse layering orders showed no significant differences. In other words, layering order is not important for solute breakthrough in layered soils with nonlinear adsorption as the dominant mechanism in one of the layers. Zhou and Selim (2001) arrived at similar conclusions for layered soil profiles when several retention mechanisms of the kinetic reversible and irreversible types were considered.

9.4.3 Langmuir Model

To illustrate that the above finding is universally valid, other solute adsorption processes of the nonlinear type were investigated. The Langmuir adsorption model is perhaps one of the most commonly used equilibrium formulations for describing various reactive solutes in porous media,

$$S/S_{\max} = \frac{\omega C}{(1 + \omega C)} \tag{9.9}$$

where $S_{\max}$ is the maximum amount sorbed and ω is the affinity coefficient. We consider here simulated columns consisting of one nonreactive layer and one reactive layer with a Langmuir-type adsorption mechanism. The simulation results are shown in Figure 9.4. The combined first- and third-type

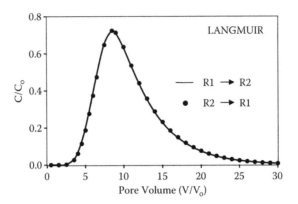

FIGURE 9.4
Simulated breakthrough results for a two-layered soil column under different layering orders (R1 → R2 and R2 → R1). Here R1 is a nonreactive layer and R2 is a reactive layer with Langmuir adsorption.

boundary condition was used at the interface between the layers. Consistent with the above finding, we found that for all parameters used in this study, the layering sequence has no effect on the BTCs when Langmuir-type adsorption was the dominant mechanism.

9.4.4 Kinetic Retention

In this section, first-, and nth-order reversible kinetics were considered as the dominant retention mechanism; namely,

$$\frac{\partial S}{\partial t} = \frac{k_1 \theta}{\rho} C - k_2 S \tag{9.10}$$

and

$$\frac{\partial S}{\partial t} = \frac{k_1 \theta}{\rho} C^n - k_2 S \tag{9.11}$$

Values of the reaction order n used were 0.3, 0.7, and 1.0. The BTCs under reverse layering orders showed a very good match regardless of the value of the nonlinear parameter n. For all cases, a good match was also realized (see Figure 9.5). We also carried out simulations where both layers were assumed reactive. Other retention mechanisms considered included irreversible reaction as well as second-order mechanism. Regardless of the retention mechanism, simulation results indicated that layered soils with reverse layering

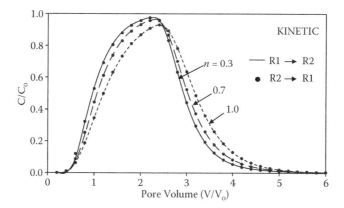

FIGURE 9.5
Simulated breakthrough results for a two-layered soil column under different layering orders (R1 → R2 and R2 → R1). Here R1 is a nonreactive layer and R2 is a reactive layer with kinetic adsorption, with $n = 0.3$, 0.7, and 1.0, respectively.

orders showed no significant differences in the BTCs. For example, in the presence of a sink term, the order of soil layers did not influence the shape or the position of the BTCs as illustrated in Figure 9.6. This finding is consistent with that of Selim, Davidson, and Rao (1977), where a similar sink term was implemented.

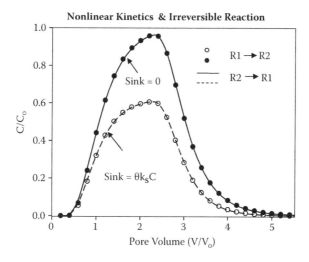

FIGURE 9.6
Simulated breakthrough results for a two-layered soil column under different layering orders (R1 → R2 and R2 → R1). Here R1 is for a nonreactive layer adsorption and R2 is a reactive layer with nth-order and irreversible sink.

9.5 Water-Unsaturated Soils

Selim, Davidson, and Rao (1977) simulated solute transport through water-unsaturated multilayered soil profiles, where a steady vertically downward water flow (q = constant) was considered. A soil profile was assumed to consist of two distinct layers, sand and clay, each having equal lengths, and was underlain by a water table at a depth $L = 100$ cm. The case where the water table was at great depth ($z \rightarrow \infty$) was also considered. When a constant flux was assumed, the steady state θ and water suction (h) distributions for a sand-clay and a clay-sand soil profile were calculated (see Figure 9.7 for the sand-clay case). Solute concentration versus pore volume of effluent (collected at 100 cm depth) for a nonreactive and reactive solute having linear (equilibrium) retention is shown in Figure 9.7. As expected, similar BTC results for the nonreactive solute for sand-clay or clay-sand soil profiles were obtained. In contrast, BTCs for the reactive solute show a distinct separation, with lower retardation factors for the soil profiles having a water table at $z = 100$ cm than at $z \rightarrow \infty$. This observation is consistent for the sand-clay as well as clay-sand profiles. Due to the higher water contents in the soil profiles where the water table was at $z = 100$ cm, the retardation factor R is less in comparison to the case where the water table was at great depth ($z \rightarrow \infty$).

If the water content distributions were considered uniform, with an average water content within each individual layer (see Figure 9.8), the problem of solute transport and retention through unsaturated multilayered soil profiles can be significantly simplified as discussed in the previous section. The open circles in Figure 9.8 are calculated results of concentration distributions for

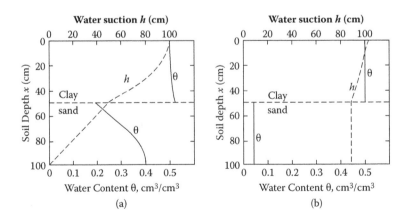

FIGURE 9.7
Simulated soil-water content θ and water suction h versus depth in a clay-sand profile having a water table (A) at 100-cm depth and (B) great depth.

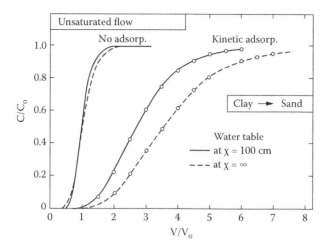

FIGURE 9.8
Simulated effluent concentration distribution for reactive and nonreactive solutes in an unsaturated clay-sand profile. Open circles are based on average water content θ for each soil layer.

the reactive and nonreactive solutes when an average water content within each layer was used. These results show that, for all unsaturated profiles considered, the use of average water contents (open circles) provided identical concentration distributions to those obtained where the actual water content distributions were used (dashed and solid lines). Thus, when a steady water flux is maintained through the profile, BTCs of reactive and nonreactive solutes at a given location in the soil profile can be predicted with average water contents within unsaturated soil layers. Based on the above results we can conclude that average microhydrologic characteristics for a soil layer can be used to describe the movement of solutes leaving a multilayered soil profile. This conclusion supports the assumption made earlier that uniform soil water content can be used to represent each soil layer in order to simplify the solute transport problem. However, such a simplifying approach was not applicable for the general case of transient water-flow conditions of unsaturated multilayered soils. As illustrated by Selim (1978), the transport of reactive, as well as nonreactive, solutes through multilayered soils, for transient water flow, was significantly influenced by the order in which the soil layers were stratified.

Solute transport in a three-layered soil profile (clay over sand over loam) is shown in Figure 9.9. In this example, we illustrate solute transport during water infiltration in an unsaturated soil (see Selim, 1978). Here application of a solute solution at the soil surface was assumed for an extended period of time, that is, continuous application. The reactivity of individual soil layers to the applied solute was assumed to follow first-order kinetics. Because of slow kinetic adsorption, the amount of solute adsorbed S continued to increase

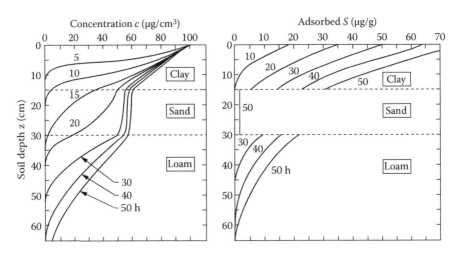

FIGURE 9.9
Solute concentration and amount of solute sorbed versus soil depth in a three-layered soil profile with first-order kinetic retention reactions.

with time in the clay layer in comparison to the other two layers. Such a slow adsorption resulted in a decrease of solute concentration with soil depth in the clay layer. In contrast, a somewhat uniform distribution was observed for the sand layer accompanied by an increase in the loam layer. Solute transport during infiltration and water redistribution was investigated for several other combinations of soil layer stratifications and the results indicated that solute transport through multilayered soil was significantly influenced by the order in which the soil layers are stratified. Another example is presented in Figures 9.10 and 9.11 from a greenhouse lysimeter study of a Windsor soil. Secondary treated wastewater was applied to the soil twice weekly for 25 weeks. Solution samples were collected and analyzed for N, and the soil-water pressure head was monitored frequently at different depths in this three-layered soil. Model predictions agreed well with pressure head data with depth and time, as well as gravimetrically determined soil water content with depth for the two soils. Model predictions of nitrate concentration shown, in Figure 9.11 with depth and time were adequate except for intermediate soil depths (30 and 45 cm). Predictions of ammonium were consistently higher than measured results for all depths (data not shown).

Examples of experimental and predicted BTC results based on miscible displacements from packed soil columns having two layers are shown in Figures 9.12 and 9.13. Predictions and experimental BTC results for tritium shown in Figure 9.13 (solid and dashed lines) for a sand-clay and clay-sand sequence indicate a good match of the experimental data. All input

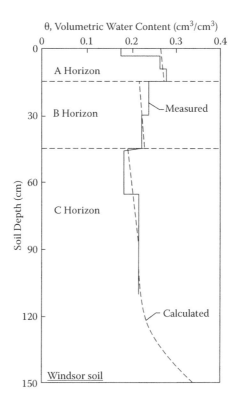

FIGURE 9.10
Measured and predicted gravimetric soil water content with depth 4 d after wastewater application for Windsor soil.

parameters were directly based on experimental measurements and support earlier findings. Zhou and Selim (2001) simulated the breakthroughs of Ca and Mg using a two-layered model for reactive solutes. The simulation results are shown in Figure 9.12 (solid and dashed lines for different layering arrangements) where the reaction mechanism for the Ca-Mg system was assumed to be governed by simple ion exchange for a binary system:

$$S = \frac{K_{12} S_T c}{1 + (K_{12} - 1)c} \tag{9.12}$$

where c is relative concentration (C/C_o), K_{12} is the selectivity coefficient (dimensionless), and S_T is the cation exchange capacity (CEC). The simulation curves agree with the experimental data, especially for the adsorption front and also exhibit tailing of the release curve for both Ca and Mg. Such

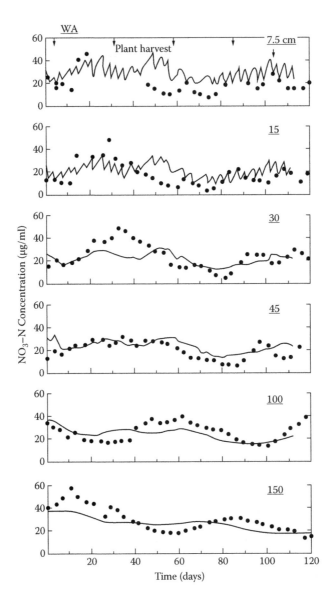

FIGURE 9.11
Measured and predicted nitrate nitrogen concentrations at various depths for Windsor soil.

tailing was not observed based on the experimental data, however. Results for nonreactive tracer solute are shown in Figure 9.13. Here experimental (symbols) and simulated (dashed and solid lines) breakthrough results for tritium in a two-layered soil column (Sharkey clay → sand, column B) under different layering orders are presented and clearly illustrate that the order of layering does not influence solute transport under steady flow conditions.

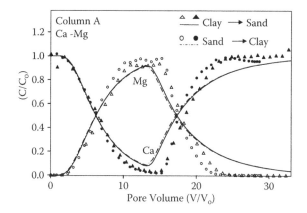

FIGURE 9.12
Experimental (symbols) and simulated (dashed and solid lines) breakthrough results for tritium in a Sharkey clay → sand column (column C) under different layering orders.

Based on the above presentation, the behavior of water and solutes in soils is dependent on the layering of the soil profile. Specifically, the physical and chemical properties of individual soil layers as well as the sequence of layering dictate the behavior of water and reactive chemicals in the soil profile. Based on mathematical analyses and laboratory experiments, solute transport is independent of the order of soil layers. This finding is applicable for reversible and irreversible solute reactions of the equilibrium and kinetic types.

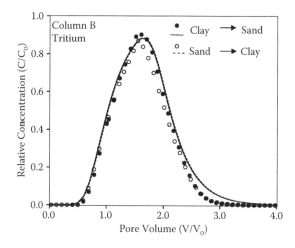

FIGURE 9.13
Experimental (symbols) and simulated (dashed and solid lines) breakthrough results for tritium in a two-layered soil column (Sharkey clay → sand, column B) under different layering orders.

These reactions include Freundlich, Langmuir, and reversible and irreversible first- and nth-order kinetics. Furthermore, simplified approaches can be used to quantify the transport of reactive chemicals in soil profile, if equilibrium reactions of the linear type are dominant for individual soil layers. Inclined soil layers are common and observed in hilly regions and sloping soils.

9.6 Mixed Media

In this section, the transport of reactive chemicals in mixed soil systems consisting of two or more porous media coupled with the second order two-site model (SOTS) and multireaction and transport model (MRTM) are presented. Applications and sensitivity analysis and examples for reactive chemical transport in mixed media are discussed.

9.6.1 Multireaction and Transport Model

The MRTM was developed to describe the time-dependent behavior of adsorption-desorption as well as transport of reactive chemicals in soils (Selim and Amacher, 1997). This multipurpose model accounts for equilibrium and kinetic reactions of both the reversible and irreversible types. The model used in this analysis can be presented in the following formulations:

$$S_e = k_e \left(\frac{\theta}{\rho} \right) C^n \tag{9.13}$$

$$\frac{\partial S_k}{\partial t} = k_1 \left(\frac{\theta}{\rho} \right) C^n - (k_2 + k_3) S_k \tag{9.14}$$

$$\frac{\partial S_i}{\partial t} = k_3 S_k \tag{9.15}$$

Here C is the concentration in solution (mg L^{-1}), S_e is the amount retained on equilibrium sites (mg kg^{-1}), S_k is the amount retained on kinetic type sites (mg kg^{-1}), and S_i is the amount retained irreversibly (mg kg^{-1}). Moreover, K_e is a dimensionless equilibrium constant, k_1 and k_2 (h^{-1}) are the forward and backward reaction rates associated with kinetic sites, respectively, k_3 (h^{-1}) is the irreversible rate coefficient associate with the kinetic sites, n is the dimensionless reaction order, θ is the soil water content (cm^3 cm^{-3}), ρ is the soil bulk density (g cm^{-3}), and t is the reaction time (h). At any time t, the total amount of atrazine retained on all sites is:

$$S_T = S_e + S_k + S_i \tag{9.16}$$

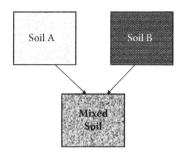

FIGURE 9.14
Mixed media as a mixture of two soils having different properties.

We now extend the MRTM formulation to a mixed soil system with two or more geologic media or soil where each medium has distinct physical, chemical, and microbiological properties (see Figure 9.14). Such parameters include θ, the soil water content ($cm^3 \, cm^{-3}$), ρ, the soil bulk density ($g \, cm^{-3}$), soil-hydraulic conductivity, and soil-water and solute retention parameters.

If one assumes that each medium competes concurrently for retention sites for a specific ion species present in the solution phase, S_T can thus be expressed as:

$$S_T = f_1 S_{T1} + f_2 S_{T2} + f_3 S_{T3} + \cdots + f_m S_{Tm} \tag{9.17}$$

Here S_{T1}, S_{T1}, S_{T1}, S_{T1}, and S_{Tm} represent the sorption capacity for geologic medium 1, 2, 3, and n, respectively. For simplicity, we assume the mixed system is composed of n media and the dimensionless parameter f (omitting subscripts) represents the fraction of medium 1, 2, 3, etc., to that of the entire mixed medium (on a mass per unit bulk volume basis). This parameter is necessary in order to account for the proportion of each geologic medium per unit bulk volume of the mixed soil system.

Based on the MRTM formulation, we can express S_T for each respective medium such that for medium 1, we have:

$$S_{T1} = S_{e1} + S_{k1} + S_{i1} \tag{9.18}$$

And the respective S_T for the mth medium can expressed as

$$S_{Tm} = S_{em} + S_{km} + S_{im} \tag{9.19}$$

Consequently, the amount sorbed by the equilibrium sites of the mixed soil

$$S_e = \sum_{l=1}^{m} f_l [S_e]_l \tag{9.20}$$

Or more explicitly we have

$$S_e = \sum_{l=1}^{m} f_l [K_e]_l \left(\frac{\theta_l}{\rho_l} \right) C^{[n]_l} \tag{9.21}$$

Similarly, the amount of solute adsorbed by the kinetic sites of the mixed soil is:

$$S_k = \sum_{l=1}^{m} f_l [S_k]_l \tag{9.22}$$

where

$$\frac{\partial [S_k]_r}{\partial t} = \sum_{l=1}^{m} f_l [k_1]_l \left(\frac{\theta_l}{\rho_l} \right) C^{[n]_l} - ([k_2]_l + [k_3]_l)[S_k]_l \tag{9.23}$$

In addition, the corresponding irreversible reactions are:

$$\frac{\partial S_{irr}}{\partial t} = \sum_{l=1}^{m} f_l [k_3]_l [S_k]_l \tag{9.24}$$

The above formulation was incorporated into the convective-dispersive equation (CDE) and solved subject to the following initial and boundary (third-type) conditions:

$$C = 0 \quad t = 0 \quad 0 < x < L \tag{9.25}$$

$$S_e = S_k = S_{irr} = 0 \quad t = 0 \quad 0 < x < L \tag{9.26}$$

$$vC_o = -D\frac{\partial C}{\partial x} + vC \quad x = 0 \quad t \le t_p \tag{9.27}$$

$$0 = -D\frac{\partial C}{\partial x} + vC \quad x = 0 \quad t > t_p \tag{9.28}$$

$$\frac{\partial C}{\partial x} = 0 \quad x = L \quad t > 0 \tag{9.29}$$

where t_p is the duration of the applied atrazine pulse (hour), L is the column length (cm), C_o is atrazine concentration in the applied pulse ($\mu g/mL$). The CDE can be solved numerically using the implicit-explicit finite-difference approximation.

9.6.2 MRTM Simulations

To illustrate the influence of mixing soils having different properties on the mobility and retention of reactive chemicals in soils or geological media, two examples are presented. The results shown are based on MRTM simulations for the respective parameters of each medium. In the first example, one assumes here a soil A which is mixed with a sand material (soil B), which is considered much less reactive compared to soil A. For simplicity, soil B is considered as a nonreactive material where all the retention coefficients such as $K_e, k_1, k_2, \ldots$, and k_{irr} are set to zero. For soil A, the retention coefficients selected for these simulations are $K_e = 1$ mg L^{-1}, $n = 0.75$, $k_1 = 0.1$ h$^-$, $k_2 = 0.01$ h^{-1}, $n = 0.5$, and $k_{irr} = 0.005$ h^{-1}. The influence of mixing a fraction (f) of the sand material on solute transport in this mixed medium is exhibited by the set of BTCs shown in Figure 9.15. When the fraction of the added sand in the mixed soil was 25% or f of 0.25, a right shift of the BTC along with an increase in concentration is observed when compared with the BTCs where no sand was added ($f = 0$). As f increases, BTC simulations indicate earlier arrival time, that is, early breakthrough of the solute in the effluent solution, along with an increased concentration of peak maxima. In addition, as f increased, a slower release or tailing during leaching is observed.

In Figure 9.15, simulations for f of 0 and 1 represent those for the respective soils A and B and illustrate the dominance of soil properties of each fraction on the resulting BTC.

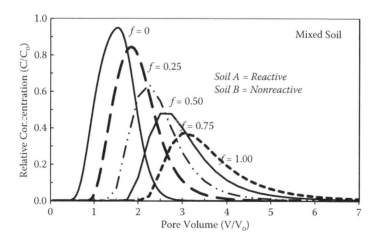

FIGURE 9.15
Miscible displacement of a solute pulse in a mixed medium made up of a mixture of two soils. Solute retention properties of soil A include irreversible, nonlinear kinetic, and nonlinear equilibrium reactions. Soil B is considered nonreactive.

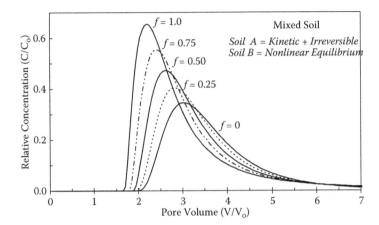

FIGURE 9.16
Miscible displacement of a solute pulse in a mixed medium made up of a mixture of two soils. Solute retention properties of soil A include irreversible, nonlinear kinetic, and nonlinear equilibrium reactions, whereas soil B has only nonlinear equilibrium reactions.

The second example is presented in Figure 9.16, which is for a somewhat similar mixed medium to that shown in Figure 9.15. In this case, the added material (soil B) is considered as a reactive material where nonequilibrium retention conditions are dominant, that is, kinetic and irreversible processes are ignored. This is illustrative of rapid or instantaneous sorption of a reactive chemical by soil B having a nonlinear sorption isotherm. Specifically, retention parameters selected for soil B are: $K_e = 1$ mg L^{-1}, $n = 0.5$ with all other parameters set to zero. The influence of mixing a fraction (f) of soil B on the mobility of solute in this mixed medium is exhibited by the set of BTCs shown in Figure 9.16. Lesser mobility of a solute in such a medium is manifested by the shapes of the BTCs when compared to a mixed soil where a nonreactive medium was added as shown in Figure 9.15. Overall, a lower BTC peak concentration maximum and spreading is observed for the case with nonequilibrium retention.

9.7 Second-Order Model (SOTS)

Now we will apply the second-order model for media of mixed soils. Specifically, the dominant retention mechanisms for each soil in the mixed system are assumed to follow that of second-order retention as previously discussed. Briefly, the second-order formulation is based on the assumption that adsorption affinities are different for the various constituents of the soil, and are represented by a system of consecutive and concurrent reactions similar to those given in Figure 9.1. The model is capable of handling concurrent and consecutive solute interactions along the lines of surface diffusion, and inter- or intra-organic matter diffusion, etc. Different sites with varied degree of affinity to solutes are analogues to concepts of solute retention via surface diffusion or intra-organic matter diffusion as discussed by Pignatello and Xing (1996), among others. A basic assumption of the second-order model is that there are a limited number of adsorption sites for solute on the soil; therefore, the reaction rates are functions of both solute concentration in solution and the availability of adsorption sites on the soil matrix. Denoting ϕ as the number of sites available for solute adsorption, the associated retention mechanisms were (Selim and Amacher, 1988):

$$S_e = K_e \theta C \phi \tag{9.30}$$

$$\frac{\partial S_k}{\partial t} = k_1 \theta C \phi - (k_2 + k_{irr}) S_k \tag{9.31}$$

$$\frac{\partial S_{irr}}{\partial t} = k_{irr} S_k \tag{9.32}$$

Here ϕ is related to the sorption capacity (S_{max}) by:

$$S_{max} = \phi + S_e + S_k \tag{9.33}$$

where ϕ and S_{max} are the unoccupied (or vacant) and total sorption sites on soil surfaces, respectively (μg solute per g soil). In addition, S_{max} was considered as an intrinsic soil property and is time invariant. The unit for K_e is cm^3 μg^{-1}, k_1 is cm^3 μg^{-1}h^{-1}, and k_2 and k_{irr} are assigned units of h^{-1}.

At equilibrium, total amounts of atrazine adsorbed on the S_e and S_k sites are:

$$S_T = S_e + S_k = S_{max} \left(\frac{\omega C}{1 + \omega C} \right) \tag{9.34}$$

Here ω [$=(K_e + K_k)\theta$] is the affinity coefficient of the combined equilibrium and kinetic adsorption, and $K_k = k_1/(k_2 + k_{irr})$. Moreover, the mass balance equation for batch reactions is then written as:

$$-\theta \frac{dC}{dt} = \rho \left(\frac{dS_e}{dt} + \frac{dS_k}{dt} \right) \tag{9.35}$$

Subject to the appropriate initial and boundary conditions, the above system of Equations 9.30 to 9.35 was solved using finite-difference approximation methods (for details see Ma and Selim, 1994).

The above formulation is now extended to mixed soils or geological media. Specifically, each soil in the mixed system has its distinct set of solute retention parameters. The simplest case is that for a mixed system composed of only two different soils with the designation r and m. If one assumes that soils r and m compete concurrently for the retention of solute present in the solution phase, S_{max} can be expressed as:

$$S_{max} = f[S_{max}]_r + (1-f)[S_{max}]_m \tag{9.36}$$

where $[S_{max}]_r$ and $[S_{max}]_m$ represent the sorption capacity for soils r and m, respectively. Here the dimensionless parameter f represents the fraction of soil r to that of the soil (on a mass per unit bulk volume basis). This parameter is necessary in order to account for the proportion of each compartment per unit bulk volume of the soil. Based on second-order formulation, we can express S_{max} for each respective compartment as:

$$[S_{max}]_r = [\phi]_r + [S_e]_r + [S_k]_r \tag{9.37}$$

$$[S_{max}]_m = [\phi]_m + [S_e]_m + [S_k]_m \tag{9.38}$$

Consequently, the amounts of solute sorbed by all equilibrium type sites are

$$S_e = f[S_e]_r + (1-f)[S_e]_m \tag{9.39}$$

or more explicitly we have:

$$S_e = \lambda\, \theta C \quad \text{where} \quad \lambda = f[K_e]_r\, \phi_r + (1-f)[K_e]_m\, \phi_m \tag{9.40}$$

Similarly, the amount of solute sorbed by the kinetic sites of both r and m soils is

$$S_k = f\,[S_k]_r + (1-f)\,[S_k]_m \tag{9.41}$$

where

$$\frac{\partial[S_k]_r}{\partial t} = [k_1]_r\, \theta C \phi_r - \{[k_2]_r + [k_{irr}]_r\}[S_k]_r \tag{9.42}$$

$$\frac{\partial[S_k]_m}{\partial t} = [k_1]_m\, \theta C \phi_m - \{[k_2]_m + [k_{irr}]_m\}[S_k]_m \tag{9.43}$$

In addition, the corresponding irreversible reactions are

$$\frac{\partial [S_{irr}]_r}{\partial t} = [k_{irr}]_r \ [S_k]_r \quad and \quad \frac{\partial [S_{irr}]_m}{\partial t} = [k_{irr}]_m \ [S_k]_m \tag{9.44}$$

The above formulation was incorporated into the convective-dispersive equation (CDE) such that:

$$R\frac{\partial C}{\partial t} = D\frac{\partial^2 C}{\partial x^2} - v\frac{\partial C}{\partial X} - f\left(\frac{\rho}{\theta}\right)\frac{\partial [S_k]_r}{\partial t} - (1-f)\left(\frac{\rho}{\theta}\right)\frac{\partial [S_k]_m}{\partial t} \tag{9.45}$$

and

$$R = 1 + \rho \lambda \tag{9.46}$$

The CDE was numerically solved using the implicit-explicit finite-difference approximation subject to the initial and boundary (third-type) conditions as discussed in previous chapters.

9.7.1 Experimental Evidence

Figures 9.17 and 9.18 are examples of the use of SOTS in a mixed soil system (Zhu, 2002). In Figure 9.17 atrazine breakthrough results from miscible displacement experiments are presented for a reference sand material mixed

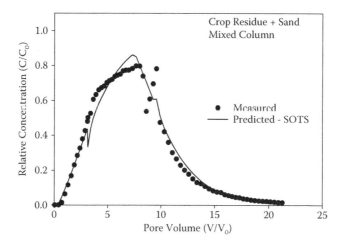

FIGURE 9.17
Measured and predicted atrazine breakthrough curves from a mixed-soil column of crop residue plus sand.

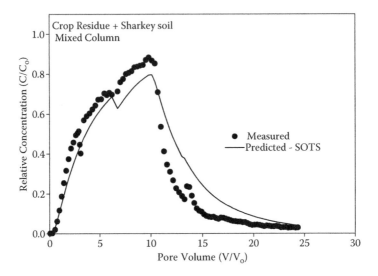

FIGURE 9.18
Measured and predicted atrazine breakthrough curves from a mixed-soil column of crop residue plus Sharkey clay soil.

with sugarcane harvest residue. In Figure 9.18, the crop residue was mixed with Sharkey clay soil (very fine, montmorillonitic nonacid, thermic Vertic Haplaquept), which was obtained from the St. Gabriel Research Station, Iberville Parish, Louisiana. The sand (acid-washed) material was used as a reference matrix where no clay or organic matter was present, where adsorption of atrazine is not expected. Furthermore, an estimate for S_{max} based on the amount of mulch incorporated into each column was made. Since it was assumed that the sand material was inert and atrazine could only be retained by the crop residue, a value of S_{max} of 98.77 µg/g was estimated for the mixed sand-crop residue column. For the Sharkey-crop residue column, estimated S_{max} of 278.83 µg/g was used. Specifically, the contributions of the crop residue to the sorptive capacity accounted for 36% (i.e., $f = 0.36$) of the total S_{max}. These estimates were based on the respective amount of crop residue mixed with the sand and Sharkey soil mixed columns.

The crop residue caused delay in atrazine breakthrough and extensive tailing as seen in Figure 9.17 for the sand-crop residue mixed column. Measured results indicated very little response in atrazine concentration to the first flow interruption at 3.1 pore volumes. In contrast, a jump in atrazine concentration in the leachate solution was observed for the second flow interruption at 9.5 pore volumes (Ma and Selim, 2005). A response to flow interruption is indicative of nonequilibrium behavior during transport caused by chemical and/or physical processes, for example, interparticle diffusion and kinetic retention (Murali and Aylmore, 1980; Reedy et al., 1996).

Predicted atrazine BTC for the mixed (sand-residue) column is shown by the solid curve shown in Figure 9.17. To obtain these predictions the retention parameters alone (K_e, k_1, k_2, and k_{irr}) were optimized and S_{max} was derived independently from batch kinetic experiments (Ma and Selim, 2005). Therefore, the SOTS model provided extremely good prediction of the overall BTC as illustrated by the solid curve in Figure 9.17. Although model calculations provided some response to the change in concentration due to flow interruption, the model failed to adequately describe the big increase in atrazine concentration during the second flow interruption (see Figure 9.17). It is possible that physical nonequilibrium played an important role during the second flow interruption, which is not accounted for by the SOTS model described above.

The SOTS model well predicted the peak position of measured atrazine but slightly underpredicted the concentration maxima for the Sharkey mixed column (see Figure 9.18). The tailing portion was somewhat overpredicted. Unlike the sand column, atrazine prediction exhibited stronger retention than that measured as illustrated by the retardation or delay of the arrival of the predicted BTC. Moreover, model prediction provided some response to the flow interruption events indicative of nonequilibrium behavior for atrazine in Sharkey soil. This model was only capable of responding to the first interruption event during sorption (see Figure 9.18). It is conceivable that physical nonequilibrium, which is not directly accounted for in SOTS, is responsible for such behavior during atrazine release. Due to the large number of parameters considered in mixed soil systems, that is, K_e, k_1, k_2, and k_{irr} for both the Sharkey clay soil and the crop residue, nonlinear optimization to improve model predictions is generally not recommended. Rather, the use of SOTS as a simplified single rather than a mixed system was examined. The solid curve in Figure 9.16 represents predicted BTCs based on a set of "weighted" model parameters. The resulting model calculations may be regarded as adequate prediction of atrazine behavior in the mixed Sharkey-crop residue column. Here the set of model parameters were "weighted averages" of the rate coefficients and were derived based on the respective contributions of the crop residue (36%) and the Sharkey clay soil (64%). Therefore, the use of weighted retention parameter values for atrazine in the SOTS model is conceivable.

References

Barry, D. A., and J. C. Parker 1987. Approximations for solute transport through porous media with flow transverse to layering. *Transport in Porous Media*, 2: 65–82.

Bosma, W. J. P., and van der Zee S. E. A. T. M. 1992. Analytical approximations for non-linear adsorbing solute transport In layered soils. *J. Contam. Hydrol*. 10: 99 118.

Guo, L., R. J. Wagenet, J. L. Hutson, and C. W. Boast. 1997. Nonequilibrium transport of reactive solutes through layered soil profiles with depth-dependent adsorption. *Environ. Sci. Technol.* 31: 2331–2338.

Leij, F. J., and J. H. Dane. 1989. Analytical and numerical solutions of the transport equation for an exchangeable solute in a layered soil. Agronomy and Soils Departmental Series No. 139. Alabama Agricultural Experimental Station, Auburn University, Alabama.

Leij, F. J., J. H. Dane, and M. Th. van Genuchten. 1991. Mathematical analysis of one-dimensional solute transport in a layered soil profile. *Soil Sci. Soc. Am. J.* 55: 944–953.

Ma, L., and H. M. Selim. 1994. Tortuosity, mean residence time and deformation of tritium breakthroughs from uniform soil columns. *Soil Sci. Soc. Am. J.* 58: 1076–1085.

Ma, L., and H. M. Selim. 2005. Predicting pesticide transport in mulch amended soils: A two compartment model. *Soil Sci. Soc. Am. J.* 69: 318–327.

Murali, V., and A. G. Aylmore. 1980. No-flow equilibrium and adsorption dynamics during ionic transport in soils. *Nature (London)* 283: 467–469.

Pignatello, J. J., and B. Xing. 1996. Mechanisms of slow sorption of organic chemicals to natural particles. *Environ. Sci. Technol.* 30: 2432–2440.

Reedy, O. C., P. M. Jardine, G. V. Wilson, and H. M. Selim. 1996. Quantifying the diffusive mass transfer of nonreactive solutes in columns of fractured saprolite using flow interruption. *Soil Sci. Soc. Am. J.* 60: 1376–1384.

Selim, H. M. 1978. Transport of reactive solutes during transient unsaturated water flow in multilayered soils. *Soil Sci.* 126: 127–135.

Selim, H. M., and M. C. Amacher. 1988. A second-order kinetic approach for modeling solute retention and transport in soils. *Water Resour. Res.* 24: 2061–2075.

Selim, H. M., and M. C. Amacher. 1997. *Reactivity and Transport of Heavy Metals in Soils.* Boca Raton, FL: CRC Press/Lewis Publishers.

Selim, H. M., J. M. Davidson, and P. S. C. Rao. 1977. Transport of reactive solutes through multilayered soils. *Soil Sci. Soc. Am. J.* 41: 3–10.

Shamir, U. Y., and D. R. F. Harleman. 1967. Dispersion in layered porous media. *Proc. Am. Soc. Civil. Eng. Hydr. Div.* 93: 237–260.

Van der Zee, S. E. A. T. M. 1994. Transport of reactive solute in soil and groundwater. In *Contamination of Groundwaters,* edited by D. C. Adriano, I. K. Iskandar, and I. P. Murarka, 27-87. Boca Raton, FL: St. Lucie Press.

van Genuchten, M. Th., and W. J. Alves. 1982. Analytical solutions of the one-dimensional convective-dispersive solute transport equation. U. S. Department of Agriculture Bulletin No. 1661.

Wu, Y. S, J. B. Kool, and P. S. Huyakorn. 1997. An analytical model for nonlinear adsorptive transport through layered soils. *Water Resour. Res.* 33: 21–29.

Zhou, L., and H. M. Selim. 2001. Solute transport in layered soils: Nonlinear kinetic reactivity. *Soil Sci. Soc. Am. J.* 65: 1056–1064.

Zhu, H. 2002. Retention and movement of reactive chemicals in soils. Ph.D. Dissertaion, Louisiana State University, Baton Rouge.

Index